Compact Textbooks in Mathematics

This textbook series presents concise introductions to current topics in mathematics and mainly addresses advanced undergraduates and master students. The concept is to offer small books covering subject matter equivalent to 2- or 3-hour lectures or seminars which are also suitable for self-study. The books provide students and teachers with new perspectives and novel approaches. They may feature examples and exercises to illustrate key concepts and applications of the theoretical contents. The series also includes textbooks specifically speaking to the needs of students from other disciplines such as physics, computer science, engineering, life sciences, finance.

- **compact:** small books presenting the relevant knowledge
- **learning made easy:** examples and exercises illustrate the application of the contents
- **useful for lecturers:** each title can serve as basis and guideline for a semester course/lecture/seminar of 2-3 hours per week.

Kenneth Shum

Measure-Theoretic Probability

With Applications to Statistics, Finance, and Engineering

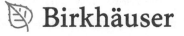 Birkhäuser

Kenneth Shum
School of Science and Engineering
The Chinese University of
Hong Kong, Shenzhen
Shenzhen, China

ISSN 2296-4568   ISSN 2296-455X (electronic)
Compact Textbooks in Mathematics
ISBN 978-3-031-49832-9   ISBN 978-3-031-49830-5 (eBook)
https://doi.org/10.1007/978-3-031-49830-5

This book is published under the imprint Birkhäuser, www.birkhauser-science.com by the registered company Springer Nature Switzerland AG
The registered company address is: Gewerbestrasse 11, 6330 Cham, Switzerland

Paper in this product is recyclable.

To the memory of my parents

Preface

Probability theory plays an increasingly important role in data science, machine learning, and many disciplines in engineering, providing mathematical tools that can model random phenomena. In data-driven applications, advanced methods from statistics are employed to analyze massive volumes of data. Bayesian inference, Markov chain Monte Carlo, and reinforcement learning, among others, are based on probability theory and stochastic processes. In finance applications, Ito integrals and stochastic differential equations are indispensable. A solid background in probability is essential for understanding state-of-the-art algorithms and developing more advanced methods.

A first course on probability theory is typically based on calculus. It is elementary and can be applied immediately to solve practical problems. However, there are some limitations in the theory. For example, random variables are usually divided into two classes, discrete and continuous, but the treatment of mixed-type random variables may be unsatisfactory. Convergence of random variables is another issue. The expectation operator in a calculus-based approach is defined in terms of the Riemann integral, which imposes some restrictions when taking limits. An even more fundamental issue is the existence of nonmeasurable sets. For example, when the underlying sample space is the unit interval in the real number line, some subsets of the interval cannot be assigned any probability.

The measure-theoretic approach to probability resolves the drawbacks of calculus-based probability mentioned in the previous paragraph. It provides a rigorous framework and serves as the language used in the exposition of more advanced methods in data science and financial engineering. There are many excellent textbooks that treat probability and measure theory comprehensively, but most of them are written for a year-long graduate-level course. With the rapid development of the areas that rely on probabilistic methods, it is beneficial to learn the basics of measure-based probability earlier at the undergraduate level.

This book is an out-growth of the course notes for an undergraduate course on probability theory, designed for students majoring in financial engineering, applied mathematics, statistics, data science, and econometrics. The students who take this course may not be interested in pure probability per se, but they need to acquire a working knowledge of probability theory so that they can read the literature and develop new methodology in their specific fields. Learning measure-based probability is like upgrading the cognitive framework of probabilistic thinking, removing bugs in the old framework, and introducing new features. Some new apps can only be supported by the new system, as they rely on the new features crucially. Hence, as a user of the probability theory, it is essential to become proficient in a rigorous foundation of probability so that more powerful theorems and results can be unlocked.

The prerequisite of this course is basic knowledge in probability and elementary concepts from real analysis. Because probability theory is built upon set theory, students are required to be familiar with set-theoretic notation.

The book is divided into 14 chapters. The first seven chapters cover the necessary mathematical tools from real analysis and measure theory. In the introductory chapter, we give some examples of singular random variables that cannot be specified using probability density function or probability mass function. We also review the notion of Riemann–Stieltjes integral in Chap. 1. In Chap. 2, we introduce sigma-fields, probability measures, and probability spaces, and in Chap. 3, we describe how to construct probability measures on the real number line. Random variables are introduced in Chap. 4. In Chap. 5, we define the notion of statistical independence of events and random variables, and discuss the Borel–Cantelli lemmas. Lebesgue integration is covered in Chaps. 6 and 7. We derive the basic convergence theorems for Lebesgue integral, the change-of-variable formula, and the connection to Riemann–Stieltjes integral.

Chapter 8 introduces the product space, coupling, optimal transport problem, and total variation distance. We also cover product measure and Fubini theorem in Chap. 8 as an important special case of coupling. In Chap. 9, we discuss the properties of moment generating functions and characteristic functions, which are essentially a type of Fourier analysis applied to probability distributions. Chapter 10 defines the various modes of convergence for random variables, such as convergence in probability as well as almost sure convergence. In Chap. 11, we prove the basic weak law and the strong law of large numbers.

Chapter 12 provides a quick introduction to functional analysis. We introduce the necessary background of Hilbert space theory to define subspace of random variables and projection of a random variable onto a subspace. Conditional expectation is introduced next in Chap. 13 as a projection operator. We also introduce filtration, martingales, and stopping times as applications of conditional expectation. The last chapter is devoted to weak convergence, Lévy's continuity theorem, and the central limit theorem.

To keep the book concise, some major results from measure theory are stated without proof, including the measure extension theorem, construction of product measures, the Fubini's theorem, the Radon–Nikodym theorem, and the Prokhorov theorem. Since these topics are well-established in the literature, we refer readers to the standard textbooks on advanced probability theory for details. Apart from these omissions, the presentation in this book is self-contained.

In addition to the theory, this book highlights several applications, including the coupon collector's problem, ball-and-bin model, the continuous mapping theorem in statistics, Monte Carlo integration in finance, data compression in information theory, and linear minimum mean-squared error estimation in signal processing.

Some chapters include Python programs that simulate the examples in the text. I believe that simulation is an integral part of studying probability theory because the ultimate goal of probability theory is to understand the random phenomena behind stochastic experiments. By running simple computer programs, we can make probability theory more concrete and gain more intuition on theoretical results.

Each chapter includes exercises with varying levels of difficulty. Some exercises are theorems used in the text that are important, including the Scheffé lemma and the π-λ theorem.

I would like to extend my gratitude to my mentors Bob Li and Ted Harris who taught me the fun and elegance of probability theory. I am also grateful to the students of MAT3280 over the years for shaping the content and the organization of this book. Special thanks to Yanyan Dong and Shuo Deng for proofreading the manuscript.

Shenzhen, China   Kenneth Shum
July, 2023

Contents

1   Beyond Discrete and Continuous Random Variables................... 1
   1.1   Discrete and Continuous Random Variables........................... 1
   1.2   Random Variables of Mixed Type and Singular Type................ 4
   1.3   Riemann–Stieltjes Integrals.................................... 8
   Problems.. 14

2   Probability Spaces.. 17
   2.1   Countable Sets.. 17
   2.2   Algebra of Events.. 19
   2.3   Measure Functions.. 23
   2.4   Borel Sets... 26
   2.5   Vitali Set... 29
   Problems.. 31

3   Lebesgue–Stieltjes Measures...................................... 33
   3.1   Pre-measure.. 33
   3.2   Stieltjes Measure Function................................. 35
   3.3   Lebesgue–Stieltjes Measures................................ 37
   3.4   Null Sets and Complete Measures............................ 45
   3.5   Uniqueness of Measure Extension............................ 46
   Problems.. 49

4   Measurable Functions and Random Variables........................ 53
   4.1   Measurable Functions....................................... 53
   4.2   Composition of Measurable Functions........................ 57
   4.3   Operations with Measurable Functions....................... 60
   4.4   Complex-Valued Random Variables............................ 65
   Problems.. 67

5   Statistical Independence... 71
   5.1   Independence of Two Random Variables....................... 71
   5.2   Independent Random Variables of Discrete Type or
         Continuous Type.. 74
   5.3   Independence of More Than Two Random Variables............. 78
   5.4   Borel–Cantelli Lemmas...................................... 81

　　　5.5　　A Model for a Sequence of Independent Random Variables......... 84
　　　Problems.. 86

6　Lebesgue Integral and Mathematical Expectation......................... 89
　　　6.1　　Simple Functions... 90
　　　6.2　　Lebesgue Integral of Nonnegative Functions.......................... 93
　　　6.3　　Lebesgue Integral of Real-Valued and Complex-Valued
　　　　　　Functions.. 97
　　　6.4　　Mathematical Expectation of Random Variable....................... 102
　　　6.5　　Application: Hat Problem and Ball-and-Bin Model.................... 104
　　　Problems.. 106

7　Properties of Lebesgue Integral and Convergence Theorems........... 109
　　　7.1　　Almost-Everywhere Equality.. 109
　　　7.2　　Fatou's Lemma and Dominated Convergence Theorem.............. 112
　　　7.3　　Application: Evaluation of Lebesgue–Stieltjes Integrals.............. 117
　　　7.4　　Push-Forward Measure and Change-of-Variable Formula........... 121
　　　7.5　　Expectation of the Product of Two Independent Random
　　　　　　Variables.. 126
　　　Problems.. 128

8　Product Space and Coupling.. 131
　　　8.1　　Coupling.. 131
　　　8.2　　Product Measure and Fubini Theorem............................... 136
　　　8.3　　Application: Monge Problem and Kantorovich Problem............. 139
　　　8.4　　Application: Total Variation Distance................................ 142
　　　Problems.. 147

9　Moment Generating Functions and Characteristic Functions........... 149
　　　9.1　　Moments and Moment Generating Functions........................ 149
　　　9.2　　Characteristic Functions.. 153
　　　　　　9.2.1　　Properties of Characteristic Functions........................ 154
　　　　　　9.2.2　　Inversion Formula.. 155
　　　　　　9.2.3　　Computing Moments from Characteristic Function........ 159
　　　Problems.. 160

10　Modes of Convergence... 163
　　　10.1　Convergence Almost Surely and Convergence in Probability....... 164
　　　10.2　Convergence in the Mean.. 167
　　　10.3　Convergence in Distribution and in Total Variation.................. 168
　　　10.4　Convergence of Random Vectors..................................... 173
　　　10.5　Application: Continuous Mapping Theorem........................... 176
　　　Problems.. 178

11　Laws of Large Numbers... 181
　　　11.1　Some Useful Bounds and Inequalities................................. 181
　　　11.2　Weak Law of Large Numbers... 183
　　　11.3　Application: Monte Carlo Integration................................. 185

	11.4	Application: Data Compression	187
	11.5	Strong Law of Large Numbers	190
	Problems		192

12 Techniques from Hilbert Space Theory 195
	12.1	L^2-Norm and Inner Product Space	195
	12.2	Closed Subspace and Projection	200
	12.3	Orthogonality Principle	202
	12.4	Application. MMSE Estimation	204
		12.4.1 Linear MMSE Estimator	204
		12.4.2 Nonlinear MMSE Estimation	206
	Problems		209

13 Conditional Expectation 211
	13.1	Expectation Conditioned on a Finite Partition	211
	13.2	Expectation Conditioned on a Sub-sigma-algebra	214
	13.3	Properties of Conditional Expectation	217
	13.4	Conditional Expectation Given a Discrete Random Variable	222
	13.5	Conditional Expectation Given a Continuous Random Variable	225
	13.6	Application: Martingale and Stopping Time	228
	Problems		234

14 Levy's Continuity Theorem and Central Limit Theorem 237
	14.1	Weak Convergence	238
	14.2	Tightness of a Sequence of Measures	241
	14.3	Prokhorov Theorem and Sequential Compactness	244
	14.4	Central Limit Theorems	246
	Problems		252

References 255

Index 257

Notation

$\mathbb{N}$	the set of integers larger than or equal to 1
$\mathbb{Z}$	the set of integers
$\mathbb{Q}$	the set of rational numbers
$\mathbb{R}$	the set of real numbers
$\mathbb{R}_{\geq 0}$	the set of nonnegative real numbers
$\mathbb{C}$	the set of complex numbers
$\mathscr{F}, \mathscr{G}$	collection of sets
$2^A, \mathscr{P}(A)$	the set of all subsets of A
$A \cup B$	union of sets
$A \cap B$	intersection of sets
A^c	set complement of a set
$A \uplus B$	disjoint union of sets
$A \setminus B$	set difference
$\{\omega\}$	singleton
$A \uplus B$	disjoint union
$|A|$	the cardinality of set A
$\sup(A)$	the smallest upper bound of a set A of real numbers
$\inf(A)$	the largest lower bound of a set A of real numbers
i.i.d.	independent and identically distributed
$\mathrm{Ber}(p)$	Bernoulli distribution
$\mathrm{Binom}(n, p)$	binomial distribution
$\mathrm{Poi}(\lambda)$	Poisson distribution
$\mathrm{Geom}(p)$	geometric distribution
$\mathrm{Unif}(a, b)$	uniform distribution between a and b
$N(\mu, \sigma^2)$	normal distribution with mean μ and variance σ^2
$N(\boldsymbol{\mu}, K)$	multivariate Gaussian distribution
$\mathrm{Exp}(\lambda)$	exponential distribution
$\Gamma(\alpha, \beta)$	Gamma distribution
$\mathrm{Beta}(a, b)$	Beta distribution
a.e, a.s	almost everywhere, almost surely
$P \ll Q$	absolute continuity for measures P with respect to Q

Beyond Discrete and Continuous Random Variables

1

Random variables are typically classified into two types: discrete and continuous. A discrete random variable takes on finitely or countably many values, while a continuous random variable can take any real number as its value. A discrete random variable can be described by a probability mass function (pmf), whereas a continuous random variable can be specified by a probability density function (pdf). By combining discrete and continuous random variables, we can construct mixed-type random variables that contain both discrete and continuous components.

Another type of random variables is known as singular random variables. These random variables are characterized by the property that their probability distribution is concentrated in a region with zero measure, making it impossible to describe them by a probability density function.

In this chapter, we will provide several examples of singular distributions. In latter chapters, we will develop a theory that can handle all of these distributions within a single framework. In the second part of this chapter, we will review the definition and basic properties of the Riemann–Stieltjes integral. By using the cumulative distribution function (cdf) of a random variable, we can compute mathematical expectation using Riemann–Stieltjes integral. This provides a unified definition of expectation for discrete, continuous, mixed-type, and singular random variables.

1.1 Discrete and Continuous Random Variables

In a first course on probability theory, we learn how to model random phenomena using two basic types of random variables. A random variable X of the first type is known as a discrete random variable. If random variable X takes nonnegative integers as its value, we can specify the distribution of X by a probability mass function $p(i) = \Pr(X = i)$ for $i = 0, 1, 2, \ldots$. Examples of discrete random variables include:

© The Author(s), under exclusive license to Springer Nature Switzerland AG 2023
K. Shum, *Measure-Theoretic Probability*, Compact Textbooks in Mathematics,
https://doi.org/10.1007/978-3-031-49830-5_1

- Bernoulli random variable, which has pmf $p(0) = 1 - p$ and $p(1) = p$ for some $p \in [0, 1]$
- Binomial random variable, whose pmf is given by $p(i) = \binom{n}{i} p^i (1 - p)^{n-i}$ for $i = 0, 1, \ldots, n$, where $p \in [0, 1]$ and n is a positive integer
- Geometric random variable, with pmf given by $p(i) = p(1 - p)^i$ for $i = 0, 1, 2, \ldots$, where p is a constant between 0 and 1
- Negative binomial random variable, which has pmf $p(i) = \binom{i+r-1}{i}(1 - p)^i p^r$, for some fixed integer $r > 0$ and real number $0 < p < 1$
- Poisson random variable, which has pmf $p(i) = \frac{\lambda^i}{i!} e^{-\lambda}$ for $i \geq 0$, where $\lambda > 0$ is a parameter of the distribution

The second type of random variables is known as continuous random variables. A continuous random variable X is described by probability density function, which is a nonnegative function $f(x)$ with integral equal to 1. The probability that X falls between a and b is computed by the integral $\int_a^b f(x)\, dx$. The followings are common distributions of continuous type:

- Uniform distribution, which is described by pdf $f(x) = (b - a)^{-1} \mathbf{1}_{[a,b]}(x)$, where $a < b$ are real numbers and $\mathbf{1}$ denotes the indicator function

$$\mathbf{1}_S(x) \triangleq \begin{cases} 1 & \text{if } x \in S \\ 0 & \text{if } x \notin S \end{cases} \tag{1.1}$$

- Gaussian distribution with mean μ and variance σ^2, which has pdf

$$f(x) = \frac{1}{\sqrt{2\pi\sigma^2}} e^{-(x-\mu)^2/(2\sigma^2)}$$

  for $x \in \mathbb{R}$
- Exponential distribution with parameter $\mu > 0$, which is specified by the pdf

$$f(x) = \mu e^{-\mu x} \mathbf{1}_{[0,\infty)}(x)$$

- Cauchy distribution, which has pdf

$$f(x) = \frac{1}{\pi(x^2 + 1)}$$

  for $x \in \mathbb{R}$
- Gamma distribution, whose pdf is given by

$$\frac{1}{\Gamma(\alpha)\beta^\alpha} x^{\alpha-1} e^{-x/\beta} \mathbf{1}_{[0,\infty)}(x), \tag{1.2}$$

where $\alpha > 0$ and $\beta > 0$ are called the shape parameter and the scale parameter, respectively, and

$$\Gamma(\alpha) \triangleq \int_0^\infty x^{\alpha-1} e^{-x}\, dx$$

is the Gamma function

The joint Gaussian distribution is a common probability model for multiple random variables. We can regard the random variables as a random column vector. If $\boldsymbol{\mu}$ is an n-dimensional column vector and K is an $n \times n$ positive definite matrix, then we define the n-dimensional Gaussian distribution by the joint distribution

$$f_{\mathbf{X}}(\mathbf{x}) = \frac{1}{(2\pi)^{n/2}\sqrt{\det(K)}} \exp\left(-\frac{1}{2}(\mathbf{x} - \boldsymbol{\mu})^T K^{-1}(\mathbf{x} - \boldsymbol{\mu})\right)$$

with $\mathbf{x}$ ranging over all vectors in $\mathbb{R}^n$. The matrix K is the covariance matrix of the random vector.

When K is the identity matrix, then the random variables in the random vector $\mathbf{X}$ are independent. For example, if $\boldsymbol{\mu} = \mathbf{0}$ is the n-dimensional zero vector and K is the $n \times n$ identity matrix, the pdf reduces to

$$f_{\mathbf{X}}(x_1, x_2, \ldots, x_n) = \frac{1}{(2\pi)^{n/2}} \exp(-(x_1^2 + x_2^2 + \cdots + x_n^2)/2).$$

This is the joint distribution of n independent standard Gaussian random variables.

If K is positive definite but not equal to the identity, the random variables are correlated. For example, when $n = 2$, $\boldsymbol{\mu} = [0, 0]^T$, and

$$K = \begin{bmatrix} \sigma_1^2 & \sigma_1\sigma_2\rho \\ \sigma_1\sigma_2\rho & \sigma_2^2 \end{bmatrix}$$

for some $\sigma_1 > 0$, $\sigma_2 > 0$, and $-1 < \rho < 1$, the joint pdf becomes

$$f_{\mathbf{X}}(x_1, x_2) = \frac{1}{2\pi\sigma_1\sigma_2\sqrt{1-\rho^2}} \exp\left(-\frac{1}{2(1-\rho^2)}\Big(\frac{x_1^2}{\sigma_1^2} + \frac{x_2^2}{\sigma_2^2} - \frac{2\rho x_1 y_1}{\sigma_1\sigma_2}\Big)\right). \quad (1.3)$$

This joint pdf describes a bivariate normal distribution with means $\mu_1 = 0$ and $\mu_2 = 0$, variances σ_1^2 and σ_2^2, and correlation coefficient ρ.

When $\rho = 1$ or $\rho = -1$, this pdf is supposed to model perfectly correlated random variables. However, in these cases, we encounter a division-by-zero error in the pdf, and hence, the pdf is not well-defined. In general, when the covariance matrix has a zero determinant, the pdf of the joint Gaussian distribution is not defined. This means that the joint Gaussian distribution is not appropriate for modeling random variables with a singular covariance matrix. In this case, the joint distribution is called a *singular Gaussian distribution*.

For example, consider the case where X_1 and X_2 have zero mean and the correlation coefficient ρ is equal to 1. The random variables X_1 and X_2 are Gaussian distributed individually, but X_2 is a deterministic linear function of X_1. Strictly speaking, the joint pdf does not exist, but we may use the Dirac delta function $\delta(x)$, which is a kind of generalized functions, to write the joint pdf as

$$f_{\mathbf{X}}(x_1, x_2) = \delta\left(x_2 - \frac{\sigma_2}{\sigma_1}x_1\right)e^{-x_1^2/(2\sigma_1^2)}/\sqrt{2\pi\sigma_1^2}.$$

1.2   Random Variables of Mixed Type and Singular Type

This section provides additional examples of probability distributions that are not discrete nor continuous. We use the term *mixed-type distribution* to refer to a distribution whose cdf can be expressed as a convex combination of a discrete random variable's cdf and a continuous random variable's cdf. A probability distribution is considered *singular* if all the probability is concentrated on a subset with measure zero.

Example 1.2.1 (Saturated Gaussian Random Variable)
Consider a saturation function $f(x)$ defined by

$$f(x) \triangleq \begin{cases} -1.5 & \text{if } x < -1.5, \\ x & \text{if } -1.5 \leq x \leq 1.5, \\ 1.5 & \text{if } x > 1.5. \end{cases}$$

Let X be a standard Gaussian random variable. The random variable $Y = f(X)$ is confined to the interval $[-1.5, 1.5]$ and is a mixture of discrete and continuous random variables. Due to the two probability masses located at the boundary of $[-1.5, 1.5]$, the cumulative distribution function $F_Y(y) \triangleq \Pr(Y \leq y)$ has jump discontinuities at $y = \pm 1.5$. The random variable Y does not have a probability density function because the cdf is not differentiable at $y = \pm 1.5$. Figure 1.1 illustrates the cdf of Y. This is an example of distribution of mixed type.

A natural question to ask in Example 1.2.1 is the conditional distribution of X given $Y = f(X)$. Suppose we can observe the value of Y and want to infer the conditional distribution of X based on the information contained in Y. However, we cannot answer this question directly using elementary probability theory because both the pdf of Y and the joint pdf of X and Y do not exist; the usual formula for conditional pdf $f_{XY}(x, y)/f_Y(y)$ does not apply. Therefore, we must resort to more advanced theory to derive the conditional distribution of X given Y.

Example 1.2.2 (Dirichlet Distribution)
Consider a positive constant β and n positive constants $\alpha_1, \ldots, \alpha_n$. For $i = 1, 2, \ldots, n$, let X_i be independent Gamma distributed random variables with shape parameter α_i and scale parameter β, which we denote by $\Gamma(\alpha_i, \beta)$. The pdf of X_i is given by (1.2).

Define $V \triangleq X_1 + X_2 + \cdots + X_n$ as the sum of these Gamma random variables. The components of the random vector

Fig. 1.1 The cumulative
distribution function in
Example 1.2.1

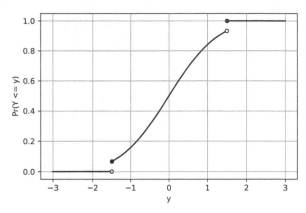

Fig. 1.2 A scatter plot of
Dirichlet distribution in
Example 1.2.2. with
parameters $\alpha_1 = \alpha_3 = 1$ and
$\alpha_2 = 2$. All sample points are
on the plane $x + y + z = 1$

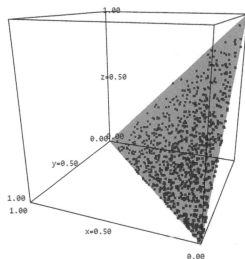

$$\mathbf{Y} = (Y_1, Y_2, \ldots, Y_n) \triangleq (X_1/V, X_2/V, \ldots, X_n/V)$$

are distributed according to the Dirichlet distribution with parameters $\alpha_1, \ldots, \alpha_n$. The random
vector $\mathbf{Y}$ lies in the region defined by $y_1 + y_2 + \cdots + y_n = 1$ and $y_i \geq 0$ for all i with probability 1.
The Dirichlet distribution is singular because this region has zero volume in $\mathbb{R}^n$. A sample scatter
plot is shown in Fig. 1.2.

The Dirichlet distribution plays a prominent role in the method of latent
Dirichlet allocation, a popular technique in natural language processing. Although
the Dirichlet distribution does not have a pdf, we can project the random variables
to an $(n-1)$-dimensional subspace and describe the probability distribution in the
lower-dimensional space. Since the n random values must sum to 1, we can consider
the first $n-1$ components only. The pdf of $Y_1, \ldots, Y_{n-1}$ is given by

$$f(y_1, \ldots, y_{n-1}) = \frac{\Gamma(\alpha_1 + \cdots + \alpha_n)}{\prod_{k=1}^{n} \Gamma(\alpha_k)} \left(\prod_{k=1}^{n-1} y_k^{\alpha_k - 1} \right) (1 - y_1 - \cdots - y_{n-1})^{\alpha_n - 1},$$

(1.4)

for $(y_1, y_2, \ldots, y_{n-1})$ with $y_1 + y_2 + \cdots + y_{n-1} \leq 1$ and $y_k \geq 0$ for $k = 1, \ldots, n - 1$. We can compute probabilities pertaining to this distribution using this lower-dimensional pdf. When $n = 2$, the pdf of Y_1 reduces to a Beta distribution,

$$f(y) = \frac{\Gamma(\alpha_1 + \alpha_2)}{\Gamma(\alpha_1)\Gamma(\alpha_2)} y^{\alpha_1 - 1} (1 - y)^{\alpha_2 - 1}$$

for $0 \leq y \leq 1$.

In Python, we can use the dirichlet function in the numpy.random module to generate a Dirichlet-distributed random vector. The following is an example of drawing a number of samples from a Dirichlet distribution with parameters $(1, 1, 2)$ using the default random number generator.

```
from numpy import random
rng = random.default_rng()     # default random number generator
X = rng.dirichlet((1, 1, 2), 8)    # draw 8 samples
print(X)  # print the random samples as an array
```

A sample run of this program yields the following output:

```
[[0.1844921  0.51764633 0.29786156]
 [0.1943196  0.18637496 0.61930544]
 [0.06687522 0.28048234 0.65264245]
 [0.50722478 0.14452905 0.34824617]
 [0.22316573 0.05919519 0.71763908]
 [0.28439455 0.22016395 0.4954415 ]
 [0.18963489 0.13846004 0.67190507]
 [0.20121907 0.28116356 0.51761736]]
```

Each row of this array represents a sample from the Dirichlet distribution. Note that the sum of the numbers in each row is equal to 1, as required by the Dirichlet distribution. In this example, we observe that the third component of each sample is often the largest, due to its high weight in the parameter vector $(1, 1, 2)$.

In each of the previous examples, there are sets with zero length or area that occur with strictly positive probability. As a result, the probability distributions associated with these examples cannot be represented by probability density functions.

We formally define a singular random variable as follows.

Definition 1.1

A real-valued random variable X is said to be *singular* if there exists a set S with length 0 such that X takes a value in S with probability 1. Similarly, a random vector $\mathbf{X}$ with values in $\mathbb{R}^n$ is called *singular* if there exists a set S with zero volume such that $\Pr(\mathbf{X} \in S) = 1$.

The Cantor distribution is another example of singular random variables.

Example 1.2.3 (Cantor Distribution)
Consider the infinite series

$$U = \sum_{i=1}^{\infty} \frac{R_i}{3^i},$$

where R_i for $i = 1, 2, 3, \ldots$ are independent discrete random variables that take value 0 or 2, each with probability 1/2. A realization of U is a number between 0 and 1. If we expand U as a 3-ary number, the digits are all equal to 0 or 2, with probability 1.

If the first digit R_1 is zero, the value of U is restricted to the interval $[0, 1/3]$, because the largest possible value given $R_1 = 0$ is $0.02222\ldots$ in base 3, which is equal to 1/3. If $R_1 = 2$, then the value of U is larger than 2/3. We thus have

$$\Pr(0 \le U \le 1/3) = \Pr(2/3 \le U \le 1) = 1/2, \text{ and } \Pr(1/3 < U < 2/3) = 0.$$

The cdf $F_U(u) \triangleq \Pr(U \le u)$ is equal to 1/2 for $u \in (1/3, 2/3)$.

By repeating the argument in the previous paragraph and considering the event $R_1 = 0$ and $R_2 = 0$ and the event $R_1 = 0$ and $R_2 = 2$, we can show that $F_U(u)$ is equal to 1/4 for $u \in (1/9, 2/9)$. Similar analysis shows that $F_U(u)$ is equal to 3/4 for $u \in (7/9, 8/9)$. The cdf of U is flat on the intervals

$$(1/3, 2/3), \ (1/9, 2/9), \ (7/9, 8/9),$$

$$(1/27, 2/27), \ (7/27, 8/27), \ (19/27, 20/27), \ (25/27, 26/27)\ldots.$$

The lengths of these intervals sum to 1. Let T be the union of all these intervals. Then, we have $\Pr(U \in T) = 0$. The set $C \triangleq [0, 1] \setminus T$ is called the *Cantor set*. This set has zero length, and we have $\Pr(U \in C) = 1$.

One can prove that the Cantor set C is uncountable. In fact, each point in the Cantor set is uniquely associated with a 3-ary number such as $0.0202202\ldots$. If we divide this number by 2, we obtain an infinite binary sequence. Conversely, an infinite binary sequence is associated with a point in the Cantor set. Because the set of all infinite binary sequences is uncountable, so is the Cantor set. The cdf of the Cantor distribution has zero slope almost everywhere, except at the uncountably many points in the Cantor set.

▶ **Notes on the Definition of Singular Distribution** According to the definition given earlier, a real-valued random variable whose range S is finite or countably infinite is classified as a singular random variable. In some textbooks, a singular distribution is further required to have zero probability at each point in the set S. With this additional requirement, a discrete random variable is not regarded as singular, but the Cantor distribution is truly singular.

In the rest of this book, however, we will not require that each point in the set S in Definition 1.1 has zero probability. Therefore, a discrete random variable is regarded as a singular random variable in this book.

The concept of singular distribution allows us to express the distribution of any real-valued random variable as the sum of a continuous part and a singular part. This result is known as the Lebesgue decomposition theorem. In practical applications, the singular part of the decomposition is often a discrete probability distribution concentrated on a set of at most countably many points.

1.3   Riemann–Stieltjes Integrals

Although there are several types of random variables, we can define the expectation of them in a unified manner using Riemann–Stieltjes integral. The Riemann–Stieltjes integral is a precursor to the more general Lebesgue–Stieltjes integral, which we will introduce in a later chapter.

Consider a closed interval $[a, b]$, with $a < b$, and let $\alpha(x)$ be a non-decreasing function on $[a, b]$. We can imagine that there is a metal wire on the interval $[a, b]$, and $\alpha(x)$ models the distribution of mass on the wire, so that $\alpha(x_2) - \alpha(x_1)$ is the mass from x_1 to x_2 ($x_1 < x_2$). In the context of probability, the function $\alpha(x)$ is the cumulative distribution function of a random variable with $\alpha(a) = 0$ and $\alpha(b) = 1$.

In the following, we let $f(x)$ be a bounded function on the interval $[a, b]$. A *partition P* of $[a, b]$ is a finite sequence of numbers between a and b, arranged in ascending order:

$$P : a = x_0 < x_1 < x_2 < \cdots < x_n = b. \tag{1.5}$$

Given a partition P of $[a, b]$ as in (1.5), we define the *upper sum* $U(P, f, \alpha)$ by

$$U(P, f, \alpha) \triangleq \sum_{k=1}^{n} M_k \cdot (\alpha(x_k) - \alpha(x_{k-1})),$$

where M_k is the maximum value of f in the subinterval $[x_{k-1}, x_k]$,

$$M_k \triangleq \sup\{f(x) : x_{k-1} \leq x \leq x_k\}.$$

Likewise, we define the *lower sum* $L(P, f, \alpha)$ by

$$L(P, f, \alpha) \triangleq \sum_{k=1}^{n} m_k \cdot (\alpha(x_k) - \alpha(x_{k-1})),$$

where m_k is the minimum value of f in the subinterval $[x_{k-1}, x_k]$,

$$m_k \triangleq \inf\{f(x) : x_{k-1} \leq x \leq x_k\}.$$

A *Riemann–Stieltjes sum* $S(P, f, \alpha)$ is defined as

$$S(P, f, \alpha) \triangleq \sum_{k=1}^{n} f(t_k) \cdot (\alpha(x_k) - \alpha(x_{k-1})),$$

where t_k is chosen arbitrarily in $[x_{k-1}, x_k]$.

Since α is a non-decreasing function, the difference $\alpha(x_k) - \alpha(x_{k-1})$ is nonnegative for all k. The Riemann–Stieltjes sum $S(P, f, \alpha)$ is thus a weighted sum of the sample function values $f(t_k)$. To simplify notation, we do not reflect the dependency on the sample points t_k's in the notation $S(P, f, \alpha)$.

We can use the following sandwiching inequalities to relate the Riemann–Stieltjes sum $S(P, f, \alpha)$ to the upper and lower sums:

$$L(P, f, \alpha) \leq S(P, f, \alpha) \leq U(P, f, \alpha), \tag{1.6}$$

which hold for any partition P.

We study the convergence of $S(P, f, \alpha)$ as the number of partitions tends to infinity by analyzing the behavior of the upper and lower sums. Note that when $\alpha(x) = x$ for all x, the Riemann–Stieltjes integral that we define below reduces to the Riemann integral. Thus, the Riemann–Stieltjes is a more flexible extension of the Riemann integral.

Definition 1.2

We continue with the notation introduced in the previous paragraphs. If the upper sum $U(P, f, \alpha)$ and lower sum $L(P, f, \alpha)$ converge to the same limit as $\max_k(x_k - x_{k-1}) \to 0$, we say that f is *Riemann–Stieltjes (RS)-integrable* on $[a, b]$ with respect to $\alpha(x)$. We denote the common limit as

$$\int_a^b f \, d\alpha \text{ or } \int_a^b f(x) \, d(\alpha(x)).$$

We use the notation $\mathscr{R}(\alpha)$ to denote the set of all RS-integrable functions on $[a, b]$ with respect to α. If f is RS-integrable on $[a, b]$ with respect to α, we write $f \in \mathscr{R}(\alpha)$.

Riemann–Stieltjes integral can be defined more generally when $\alpha(x)$ belongs to the more general class of bounded-variation functions. In probability theory, the cdf of random variables are non-decreasing, and so integration with respect to non-decreasing functions is sufficient for our purposes.

From (1.6), we see that if $f \in \mathscr{R}(\alpha)$,

$$\lim L(P, f, \alpha) = \lim S(P, f, \alpha) = \lim U(P, f, \alpha)$$

as $\max_k(x_k - x_{k-1}) \to 0$, and all three limits converge to the same value, which is equal to $\int_a^b f \, d\alpha$. The next example is a special case of Riemann–Stieltjes integrals.

Example 1.3.1 (Riemann–Stieltjes Integral When α Is a Step Function)
Suppose $\alpha(x)$ is a step function with a single discontinuity at c,

$$\alpha(x) = \begin{cases} u_1 & \text{if } x < c, \\ u_2 & \text{if } x \geq c, \end{cases}$$

for some $u_1 < u_2$. Assume that $f(x)$ is continuous at $x = c$. Then

$$\int_a^b f \, d\alpha = f(c)(u_2 - u_1).$$

The reason for this is that in the Riemann–Stieltjes sum, the only term that can be nonzero is $f(t_k)(u_2 - u_1)$, where t_k is a number in the subinterval containing c. As the size of the subintervals in the partition approaches zero, t_k approaches c, and hence, $f(t_k)$ approaches $f(c)$.

Theorem 1.1

Suppose f is bounded on $[a, b]$ and there are finitely many points $c_1, c_2, \ldots, c_n$ in $[a, b]$ such that f is discontinuous at each c_i. If the function α is continuous on each of these discontinuity points of f, then $f \in \mathcal{R}(\alpha)$.

See [7, Theorem 6.10] for a proof of Theorem 1.1.

Theorem 1.2

1. *If $f \in \mathcal{R}(\alpha)$ and $g \in \mathcal{R}(\alpha)$, then $f + g \in \mathcal{R}(\alpha)$ and*

$$\int_a^b f + g \, d\alpha = \int_a^b f \, d\alpha + \int_a^b g \, d\alpha.$$

2. *If $f \in \mathcal{R}(\alpha)$, then $cf \in \mathcal{R}(\alpha)$ for any constant c and*

$$\int_a^b cf \, d\alpha = c \int_a^b f \, d\alpha.$$

3. *If $f, g \in \mathcal{R}(\alpha)$ and $f(x) \leq g(x)$ for all $x \in [a, b]$, then*

$$\int_a^b f \, d\alpha \leq \int_a^b g \, d\alpha.$$

4. *Suppose $a < c < b$. If $f \in \mathcal{R}(\alpha)$ on $[a, c]$ and $f \in \mathcal{R}(\alpha)$ on $[c, b]$, then f is RS-integrable on $[a, b]$, and*

$$\int_a^b f \, d\alpha = \int_a^c f \, d\alpha + \int_c^b f \, d\alpha.$$

These properties are all analogous to the properties of Riemann integrals, and hence, the proofs are omitted. The following property concerns the effect of changing the function $\alpha(x)$ on the Riemann–Stieltjes integral.

Theorem 1.3
If $f \in \mathscr{R}(\alpha_1)$ and $f \in \mathscr{R}(\alpha_2)$ for two non-decreasing functions α_1 and α_2, then $f \in \mathscr{R}(\alpha_1 + \alpha_2)$ and

$$\int_a^b f\, d(\alpha_1 + \alpha_2) = \int_a^b f\, d\alpha_1 + \int_a^b f\, d\alpha_2.$$

Proof It is clear that $\alpha_1 + \alpha_2$ is a non-decreasing function. Let P be a partition of the interval $[a, b]$ as in (1.5) and choose any point $t_k \in [x_{k-1}, x_k]$. The Riemann–Stieltjes sum of f with respect to $\alpha_1 + \alpha_2$ can be decomposed as

$$S(P, f, \alpha_1 + \alpha_2) = \sum_{k=1}^n f(t_k)\big[\alpha_1(x_k) + \alpha_2(x_k) - (\alpha_1(x_{k-1}) + \alpha_2(x_{k-1}))\big]$$

$$= \sum_{k=1}^n f(t_k)(\alpha_1(x_k) - \alpha_1(x_{k-1})) + \sum_{k=1}^n f(t_k)(\alpha_2(x_k) - \alpha_2(x_{k-1})).$$

As $n \to \infty$ and $\max_k(x_k - x_{k-1}) \to 0$, we have

$$S(P, f, \alpha_1 + \alpha_2) \to \int_a^b f\, d\alpha_1 + \int_a^b f\, d\alpha_2.$$

∎

Theorem 1.4
Suppose f is a continuous function on $[a, b]$ and let α be a non-decreasing function on $[a, b]$ with a continuous derivative $\alpha'(x)$. If $f \in \mathscr{R}(\alpha)$, then the function $f\alpha'$ is Riemann integrable on $[a, b]$, and we have

$$\int_a^b f\, d\alpha = \int_a^b f(\alpha)\alpha'(x)\, dx.$$

Proof Fix a partition P as in (1.5). Because α is differentiable, by the mean value theorem, we can find a point $t_k \in [x_{k-1}, x_k]$ such that

$$\alpha(x_k) - \alpha(x_{k-1}) = \alpha'(t_k)(x_k - x_{k-1}).$$

We can write the Riemann–Stieltjes sum $S(P, f, \alpha)$ as

$$S(P, f, \alpha) = \sum_{k=1}^{n} f(t_k)\alpha'(t_k)(x_k - x_{k-1}).$$

Since f and α' are continuous, their product is also continuous. We can treat the above sum as a Riemann sum and use it to approximate the Riemann integral $\int_a^b f(x)\alpha'(x)\,dx$.   ∎

About the Assumption of $\alpha(x)$ in Theorem 1.4

The condition "$\alpha'(x)$ is continuous on [a,b]" in the previous theorem can be relaxed to "$\alpha'(x)$ is Riemann integrable on $[a, b]$". See [7, Theorem 6.17]. This result has important implications for the theory of Riemann–Stieltjes integration, as it allows for a wider class of functions to be integrated using the Riemann–Stieltjes integral.

Using Riemann–Stieltjes integral, we can define the expectation of a random variable in a unified manner.

Definition 1.3

Let X be a random variable, and let $F_X(x) = \Pr(X \le x)$ be its cdf. We define the *expectation* and *variance* of X as follows:

$$E[X] \triangleq \int x\,dF_X(x),$$

$$\mathrm{Var}(X) \triangleq \int (x - E[X])^2\,dF_X(x).$$

In this definition, the expectation of X is the weighted average of its possible values, where the weights are given by the probabilities in the cdf $F_X(x)$. The variance of X measures how much its values typically deviate from their expected value and is also based on the probabilities in the cdf $F_X(x)$. The definition of expectation and variance of a random variable X applies to random variables of discrete type, continuous type, and singular type. If the cdf is differentiable with derivative $f_X(x)$, we can use Theorem 1.4 to simplify the calculation of the expectation:

$$E[X] = \int x f_X(x)\,dx.$$

In this case, the expectation is the integral of the product of x and the pdf $f_X(x)$, which is an ordinary Riemann integral. This allows us to compute the expectation using standard techniques from calculus.

Suppose X is a discrete random variable that takes values in $c_1, c_2, c_3, \ldots, c_m$ with probabilities $\Pr(X = c_i) = p_i$, for $i = 1, 2, \ldots, m$. We can write the cdf $F_X(x)$ of X as

$$F_X(x) = \sum_{i=1}^{m} p_i \cdot \mathbf{1}_{[c_i, \infty)}(x),$$

where $\mathbf{1}_{[c, \infty)}(x)$ is the step function defined in (1.1).

By applying Theorem 1.3 several times, we can express the expectation of a discrete random variable X as a finite sum. Let $a < c_1$ and $b > c_m$. The expectation of X can be written as

$$\int_a^b x \, dF_X(x) = \sum_{i=1}^{m} \int_a^b x \, d(p_i \mathbf{1}_{[c_i, \infty)}) = \sum_{i=1}^{m} c_i p_i.$$

This recovers the formula that computes the expectation of a discrete random variable X using its probability mass function.

Using the Riemann–Stieltjes integral, we can also define the expectation of a Cantor distributed random variable. Recall that the Cantor function $F_U(u)$ is a non-decreasing function that increases from $F_U(0) = 0$ to $F_U(1) = 1$. We can define the expectation of a Cantor distributed random variable U as

$$E[U] = \int_0^1 u \, dF_U(u).$$

One can show that the expectation of the Cantor distribution is $1/2$ and the variance is $1/8$. This formula extends the concept of expectation to a singular continuous distribution that does not have a probability density function or a probability mass function.

For random variable whose range is not bounded, we can extend the Riemann–Stieltjes integral to an improper integral to compute its expectation.

Definition 1.4

We define an improper Riemann–Stieltjes integral of a function f with respect to a non-decreasing function α by the double limit

$$\int_{-\infty}^{\infty} f \, d\alpha \triangleq \lim_{\substack{a \to -\infty \\ b \to \infty}} \int_a^b f \, d\alpha$$

provided that the limit exists and is finite. In this case, we say that the improper integral converges.

Improper Riemann–Stieltjes integrals are useful for computing the expectation of random variables whose range is not bounded. However, we note that not all functions have well-defined improper integrals, and care must be taken in computing them.

Example 1.3.2 (Computing the Mean of a Random Variable of Mixed Type)
Consider the random variable Y in Example 1.2.1. Let p be the probability that a standard normal random variable is larger than 1.5. We can decompose the cdf of Y into a discrete part and a continuous part as follows.

The first part corresponds to the discrete part of Y and is given by

$$\alpha_1(x) \triangleq p\mathbf{1}_{[-1.5,\infty)}(x) + p\mathbf{1}_{[1.5,\infty)}(x).$$

The second part corresponds to the continuous part of Y and is given by

$$\alpha_2(x) \triangleq \begin{cases} 0 & \text{if } x < -1.5, \\ (1-2p)\int_{-1.5}^{x} \frac{1}{\sqrt{2\pi}}e^{-t^2/2}\,dt & \text{for } -1.5 \le x < 1.5, \\ 1-2p & \text{if } x \ge 1.5. \end{cases}$$

By Theorem 1.3, we can compute the expectation of Y as

$$E[Y] = \int_{-\infty}^{\infty} y\,d\alpha_1(y) + \int_{-\infty}^{\infty} y\,d\alpha_2(y).$$

Since both integrals are zero by symmetry, the expectation of Y is zero.

In this section, we have seen how to give a unified treatment of defining the mathematical expectation of random variables of various types, including discrete random variables, continuous random variables, and random variables of mixed type. By using the concept of the cumulative distribution function, we can decompose the expectation of a random variable into a sum of its discrete and continuous parts and then compute each part separately.

However, the Riemann–Stieltjes integral has some limitations compared to the Lebesgue integral that we are going to introduce in a latter chapter. While the Riemann–Stieltjes integral can integrate a wide class of functions, it is not powerful enough to integrate certain types of functions that arise in advanced theory. For example, it cannot integrate the indicator function of the rational numbers.

In contrast, the Lebesgue integral is more versatile but is also more abstract. It requires a deeper understanding of measure theory to unfold its power.

Problems

1.1. Let X and Y be independent jointly Gaussian random variables, each with mean zero and variance σ^2. Show that $X^2 + Y^2$ is exponentially distributed with mean $2\sigma^2$.

1.2. We generate a random variable X according to the Poisson distribution with pmf $p(X = n) = \frac{\lambda^n}{n!}e^{-\lambda}$, for integer $n \geq 0$. Then we generate X independent Bernoulli random variables $Y_1, Y_2, \ldots, Y_X$, each with success probability p. Show that the random sum $Y_1 + Y_2 + \cdots + Y_X$ is Poisson distributed with mean $p\lambda$.

1.3. Suppose X and Y are independent random variables with Gamma distributions having the same scale parameter:

(a) Prove that the sum $X + Y$ is also Gamma distributed with the same scale parameter.
(b) Prove that $X/(X + Y) \sim \text{Beta}(\alpha_1, \alpha_2)$.
(c) Show that the Beta distribution $\text{Beta}(\alpha_1, \alpha_2)$ has expected value $\frac{\alpha_1}{\alpha_1 + \alpha_2}$.

1.4. Extend the answer of the previous exercise to n independent Gamma distributed random variables, and derive the pdf of the Dirichlet distribution in (1.4).

1.5. Suppose X and Y are zero-mean, independent, and jointly Gaussian random variables with variance $E[X^2] = E[Y^2] = 1$. Let U and V be a linear transformation of X and Y computed by

$$\begin{bmatrix} U \\ V \end{bmatrix} = \begin{bmatrix} a & b \\ b & a \end{bmatrix} \begin{bmatrix} X \\ Y \end{bmatrix},$$

where a and b are constants:

(a) Find all combinations of a and b such that the random vector (U, V) is singular.
(b) Write down the joint pdf of (U, V) when it is not singular.
(c) How would you describe (U, V) in the singular case?

1.6. We roll a fair die and denote the value minus 1 as X, so that X is a discrete uniform random variable taking values in $\{0, 1, 2, 3, 4, 5\}$. Independently, we generate a continuous uniform random variable U between 0 and 1. Let Y be the product XU. The random variable Y can be described as follows. When $X = 0$, Y is equal to a constant 0. When $X = 1$, Y takes on a uniform distribution between 0 and 1. When $X = 2$, Y takes on a uniform distribution between 0 and 2, and so on:

(a) Draw the cumulative distribution function of Y.
(b) Compute the expectation of Y.

1.7. Define the floor function $\lfloor x \rfloor$ of a real number x by the largest integer less than or equal to x. Let U be a random variable uniformly distributed between 0 and 2, and $V = \lfloor U \rfloor$. The joint cumulative distributions of U and V are defined as $F_{UV}(u, v) = \Pr(U \leq u, V \leq v)$. Describe the joint distribution of U and V by computing $F_{UV}(u, v)$ for all $u \in \mathbb{R}$ and $v \in \mathbb{R}$.

1.8. Evaluate the Riemann–Stieltjes integrals $\int_a^b f\, d\alpha$ with a, b, $f(x)$, and $\alpha(x)$ given below:

(a) $[a, b] = [0, 10]$, $f(x) = x^2$, $\alpha(x) = \lfloor x \rfloor$.
(b) $[a, b] = [0, 2]$, $f(x) = \sqrt{x}$, $\alpha(x) = \lfloor x \rfloor + x$.

($\lfloor x \rfloor$ is the floor function as in the previous exercise.)

1.9. Suppose $f(x)$ is a non-decreasing function on $[a, b]$ and $f(x) \in \mathscr{R}(\alpha)$ for some non-decreasing function $\alpha(x)$. Furthermore, suppose that $\alpha(x) \in \mathscr{R}(f)$. The objective of this question is to prove the integration-by-parts formula for Riemann–Stieltjes integral:

$$\int_a^b \alpha\, df = f(b)\alpha(b) - f(a)\alpha(a) - \int_a^b f\, d\alpha. \qquad (1.7)$$

Fix a partition

$$P : a = x_0 < x_1 < \cdots < x_n = b$$

of $[a, b]$, and let t_k be a number between x_{k-1} and x_k, for $k = 1, 2, \ldots, n$.

(a) Show that

$$f(b)\alpha(b) - f(a)\alpha(a) = \sum_{k=1}^n f(x_k)\alpha(x_k) - \sum_{k=1}^n f(x_{k-1})\alpha(x_{k-1}).$$

(b) Use part (a) to show that

$$f(b)\alpha(b) - f(a)\alpha(a) - S(P, \alpha, f)$$

can be expressed as a Riemann–Stieltjes sum $S(P', f, \alpha)$ with partition

$$P' : a = x_0 \leq t_1 \leq x_1 \leq t_2 \leq x_2 \leq \cdots \leq t_n \leq x_n = b.$$

(c) Prove the integration-by-parts formula in (1.7).

1.10. Let X be a random variable whose value is nonnegative real numbers. Suppose the cdf $F_X(x) = P(X \leq x)$ satisfies $x(1 - F(x)) \to 0$ as $x \to \infty$. Derive the following formula for the expectation of X:

$$E[X] = \int_0^\infty (1 - F_X(x))\, dx.$$

Probability Spaces

2

A probability space is a mathematical model used to represent a random experiment. It consists of a sample space, a σ-algebra, and a probability measure. The sample space is the set of all possible outcomes of the experiment. During the random experiment, one outcome is selected, which represents a realization of the experiment. We may not be able to observe the chosen outcome directly, but only whether it belongs to a subset of the sample space. We may also be interested in whether two events occur simultaneously or whether at least one of them happens. The theory of probability must be rich enough to combine events using basic set operations.

To model the random experiment, we assign probabilities to the events in a consistent manner, which is captured by the probability measure function. The sample space can be a metric space or an abstract space. In this chapter, we first introduce measure and probability spaces in general and go through some basic properties. Then, we focus on sample spaces in which we can talk about open sets, such as Euclidean space and metric space, and discuss the Borel algebra generated by the open sets. This will allow us to study probability measures on more specific sample spaces.

2.1   Countable Sets

A set is countable if the elements can be listed sequentially.

Definition 2.1

Two sets are said to have the same size, or cardinality, if there exists a bijection between them. A set is said to be *countable* if it can be mapped bijectively to the set of natural numbers. An infinite set that is not countable is called *uncountable*.

© The Author(s), under exclusive license to Springer Nature Switzerland AG 2023   17
K. Shum, *Measure-Theoretic Probability*, Compact Textbooks in Mathematics,
https://doi.org/10.1007/978-3-031-49830-5_2

Example 2.1.1 (Countable Sets)

- The set of positive even integers is a subset of natural numbers $\mathbb{N}$. Nevertheless, it has the same cardinality as $\mathbb{N}$ and is therefore countable. We have an explicit bijection $f(x) = 2x$ from $\mathbb{N}$ onto the positive even integers.
- The set of integers $\mathbb{Z}$ are countable because we can list the integers as follows: $0, 1, -1, 2, -2, 3, -3, \ldots$.
- An infinite subset of a countable set is countable.
- The union of two countable sets is countable. In particular, if an uncountable set is the union of two sets, then one of these sets must be uncountable.
- The Cartesian product of two countable sets is countable.
- The set of positive rational numbers is countable because it can be put into one-to-one correspondence with the subset

$$\{(a, b) \in \mathbb{N} \times \mathbb{N} : \gcd(a, b) = 1\},$$

which is a subset of $\mathbb{N} \times \mathbb{N}$. The set of rational numbers between 0 and 1 is also countable because it is a subset of the positive rational numbers.

▶ **Convention** Some people extend the notion of countable set to include finite sets as well. In this book, however, we will use the term "countable" exclusively for infinite sets that are in bijection with the set of natural numbers. To refer to sets that are either finite or countable, we will use the term *at most countable*. We say that a set is *infinitely countable* if we want to emphasize that the set has infinite cardinality.

Example 2.1.2 (Cantor's Diagonal Argument)
Using Cantor's diagonal argument, we can show that the set of real numbers in the unit interval $[0, 1]$ is uncountable. To do this, we first consider the set S of infinite binary sequences. Each such sequence has the form $(b_1, b_2, b_3, b_4, \ldots)$, with $b_k \in \{0, 1\}$ for all $k \geq 1$.

To prove that the set S is uncountable, we assume for the sake of contradiction that S is countable and hence can be enumerated as a list containing all the infinite binary sequences. Then we can enumerate the binary sequences in a list. This can be viewed as an infinite array of binary numbers. We denote the j-th bit in the i-th sequence by b_{ij}. We assume that the list is complete.

We then construct another binary sequence s^* by extracting and flipping the diagonal entries in the array of bits, starting with the first bit of the first sequence, the second bit of the second sequence, and so on. The i-th bit in s^* is defined as $1 - b_{ii}$. Since s^* differs with each infinite sequence in the list in at least one position, it is not equal to any one of them. This contradicts the assumption that we can exhaustively enumerate all the infinite binary sequences as a list.

For example, suppose we have a list of binary sequences as follows:

$$s_1 = \mathbf{1}011011000111001\ldots$$

$$s_2 = 1\mathbf{1}11000001101101\ldots$$

$$s_3 = 00\mathbf{1}0000001001101\ldots$$

$$s_4 = 101\mathbf{0}001110000111\ldots$$

$$\vdots = \vdots$$

Each sequence in the list has infinitely many digits, and the i-th bit in the i-th number in the list is displayed in boldface. In this case, we have $s^* = 0001\ldots$, which is not in the list.

Next, we associate a binary sequence $(b_1, b_2, b_3, \ldots,)$ with a real number by the formula

$$\sum_{j=1}^{\infty} b_j 2^{-j}.$$

However, this map is not necessarily injective, as different sequences can correspond to the same real number. For example, $0.10000\ldots$ and $0.011111\ldots$ are both equal to $1/2$. To obtain an injection, we consider only the binary sequences that do not end with infinitely many trailing zeros. The removed sequences correspond to the dyadic rationals in $[0, 1]$, which are countable. Since we have removed only a countable set from S, the resulting set of binary sequences that do not end with infinitely trailing zeros is uncountable. This means that there exists a one-to-one function from an uncountable set to the interval $[0, 1]$. Hence, $[0, 1]$ is uncountable.

One can prove that the set of all real numbers and the interval $(0, 1)$ has the same cardinality. For example, we can use the arctangent function $\frac{1}{\pi} \tan^{-1}(x) + 0.5$. Therefore, the set of real numbers is uncountable.

2.2   Algebra of Events

The algebra of events is akin to a language, with a larger algebra representing a more extensive vocabulary. A larger algebra provides more events to reason about the outcomes of a random experiment, similar to how a richer language provides more means to communicate ideas. We use the following example to motivate the notion of an algebra of events.

Example 2.2.1 (Information Embedded in the Partition of Sample Space)
Alice generates two random bits of information. Whether the bits are uniformly distributed or statistically independent is not relevant in this example. We are interested in the possible events that Alice can observe. Naturally, Alice can take the sample space

$$\Omega = \{00, 01, 10, 11\}$$

as the set of all outcomes. Any subset of Ω is a valid event that Alice can observe.
Alice has two friends Bob and Cindy. Bob wants to determine the number of ones that appear among the two random bits, and Cindy wants to know whether the two bits are the same.
Upon observing the generated bits, Alice will inform Bob of an integer between 0 and 2. Bob will know that one of the following three events has occurred:

$$\{00\}, \{01, 10\}, \{11\}.$$

Alice will inform Cindy that one of the events

$$\{00, 11\}, \{10, 01\}$$

occurred. In fact, Alice does not need to inform Cindy herself. Alice may ask Bob to tell Cindy whether the two bits are the same, as Bob has enough information to do so. Mathematically speaking, this is because Cindy's partition of the sample space is coarser than Bob's partition, meaning that Bob has more detailed information about the possible outcomes of the random bits.

This example illustrates that in some cases, we may only be interested in certain subsets of the sample space. In order to work with these subsets in a rigorous mathematical way, we need to define a collection of events that is closed under the usual set operations. This leads us to the notion of an algebra of events.

Definition 2.2

A collection of subsets $\mathcal{F}$ of a sample space Ω is called an *algebra* or a *field* if it satisfies the following axioms:

1. The entire sample space Ω is an element of the collection: $\Omega \in \mathcal{F}$.
2. The collection is closed under finite unions: $A \in \mathcal{F}$ and $B \in \mathcal{F}$ imply $A \cup B \in \mathcal{F}$.
3. The collection is closed under complement: If $A \in \mathcal{F}$, then its complement A^c (taken relative to Ω) is in $\mathcal{F}$.

A set in $\mathcal{F}$ is called an *event*.

In this book, we will use the notation "algebra" and "field" interchangeably.

Using de Morgan's law, we can deduce that an algebra is closed under taking intersections as well as unions. In fact, we can perform finitely many set operations to the subsets of an algebra, and the resulting set will be an event in the algebra.

A trivial example of an algebra is $\mathcal{F} = \{\emptyset, \Omega\}$. In this algebra, the only subsets that are considered events are the empty set and the entire sample space Ω. At the other end of the spectrum, we have the power set $\mathcal{F} = \mathcal{P}(\Omega)$, which is the set of all possible subsets of Ω. In this case, every subset of Ω is considered an event.

In Example 2.2.1, Alice has full information, so she can consider the power set $\mathcal{F}_A = \mathcal{P}(\Omega)$ as the algebra of events. For Bob, the algebra of events is given by

$$\mathcal{F}_B = \{\emptyset, \{00\}, \{11\}, \{01, 10\}, \{00, 11\}, \{00, 01, 10\}, \{01, 10, 11\}, \Omega\}.$$

This algebra includes all possible events that Bob can observe, based on the information that Alice has provided him. We can see that $\mathcal{F}_A \supset \mathcal{F}_B$, which makes sense since Alice has more information than Bob.

For Cindy, the algebra of events is given by

$$\mathcal{F}_C = \{\emptyset, \{10, 01\}, \{00, 11\}, \Omega\}.$$

This algebra includes all possible events that Cindy can observe. We can see that $\mathcal{F}_A \supset \mathcal{F}_B \supset \mathcal{F}_C$, as Bob has more detailed information than Cindy.

In practice, we may encounter situations where we have infinitely many events, and we want to compute the probability that all of them occur simultaneously. In probability theory, this type of operation is allowed as long as the events involved are at most countable.

> **Definition 2.3**
>
> A collection of subsets $\mathscr{F}$ of a sample space Ω is called a σ-*algebra* or a σ-*field* if it satisfies the following axioms:
>
> 1. $\Omega \in \mathscr{F}$.
> 2. If $(A_i)_{i=1}^{\infty}$ is a sequence of sets in $\mathscr{F}$, then $\cup_{i=1}^{\infty} A_i \in \mathscr{F}$.
> 3. If $A \in \mathscr{F}$, then $A^c \in \mathscr{F}$.
>
> A set in $\mathscr{F}$ is called $\mathscr{F}$-*measurable*, or simply *measurable* if $\mathscr{F}$ is understood from the context. We say that $(\Omega, \mathscr{F})$ is a *measurable space* if $\mathscr{F}$ is a σ-algebra of Ω.
>
> If $\mathscr{F}$ is a σ-field and $\mathscr{G}$ is a sub-collection of $\mathscr{F}$, we say that $\mathscr{G}$ is a *sub-σ-field* or a *sub-σ-algebra* of $\mathscr{F}$ if $\mathscr{G}$ is also a σ-field.

Set Notation
Let $\mathcal{I}$ be an index set, which may be uncountably infinite, and A_i be a subset of a universal set U, for $i \in \mathcal{I}$. We denote the union of all A_i over all $i \in \mathcal{I}$ by

$$\bigcup_{i \in \mathcal{I}} A_i \triangleq \{x \in U : x \in A_i \text{ for some } i \in \mathcal{I}\},$$

which is the set of all $x \in U$ such that x belongs to A_i for some index i in $\mathcal{I}$. If $\mathcal{I}$ is countable, say $\mathcal{I} = \mathbb{N}$, we may use the notation

$$\bigcup_{i=1}^{\infty} A_i \triangleq \{x \in U : x \in A_i \text{ for some } i \in \mathbb{N}\}$$

to represent the union of A_i over $i \in \mathbb{N}$. Similarly, the intersection of infinitely many sets can be defined using analogous notation.

It is important to note that these are mere notational conventions, and no limit process is involved in taking the union or intersection of infinitely many sets. In the special case where the sets A_i are mutually disjoint, i.e., $A_i \cap A_j = \emptyset$ whenever $i \neq j$, we may use the notation $\biguplus_{i=1}^{\infty} A_i$ to emphasize that the union is disjoint.

It is clear that a σ-algebra is always an algebra. When the sample space Ω is finite, a σ-algebra is the same as an algebra. However, when the sample space Ω is infinite, it is possible to construct σ-algebra that is not an algebra.

Example 2.2.2 (Co-finite Algebra)
Consider the set $\Omega = \mathbb{N}$, and let $\mathscr{G}$ be the collection of subsets of $\mathbb{N}$ that are finite or co-finite, i.e., $A \in \mathscr{G}$ if and only if $|A| < \infty$ or $|A^c| < \infty$. For example, $\{1, 2, 3\}$ is in $\mathscr{G}$ because it is finite, and $\mathbb{N} \setminus \{2, 3, 5, 7\}$ is in $\mathscr{G}$ because its complement is finite. We can verify that $\mathscr{G}$ is an algebra. However, $\mathscr{G}$ is not a σ-algebra because the sequence of events $\{2\}, \{4\}, \{6\}, \{8\}, \{10\}, \ldots$ is in $\mathscr{G}$, but their union is not.

There are two useful set constructions called liminf and limsup.

Definition 2.4

Let $E_1, E_2, \ldots$ be an arbitrary sequence of subsets in a set Ω. The *limit inferior* and *limit superior* of $(E_i)_{i \geq 1}$ are defined, respectively, as

$$\liminf_{i \to \infty} E_i \triangleq \bigcup_{j=1}^{\infty} \bigcap_{k \geq j} E_k, \qquad \limsup_{i \to \infty} E_i \triangleq \bigcap_{j=1}^{\infty} \bigcup_{k \geq j} E_k.$$

In general, we have the following set inclusion:

$$\liminf_{i \to \infty} E_i \subseteq \limsup_{i \to \infty} E_i.$$

If equality holds, we say that the *limit of* $(E_i)_{i \geq 1}$ exists and is defined as $\liminf_i E_i$ or $\limsup_i E_i$.

We remark that the liminf and limsup of a sequence of sets are always well-defined as set-theoretic constructions. However, if the sequence of sets $(E_i)_{i \geq 1}$ is $\mathcal{F}$-measurable, then the limit inferior and limit superior of this sequence are also $\mathcal{F}$-measurable. This is because their definitions rely only on the union and intersection of countably many events, which are themselves $\mathcal{F}$-measurable.

The following theorem provides an alternate characterization of liminf and limsup.

Theorem 2.1 (Definition 2.4)

Let $(E_i)_{i \geq 1}$ be a sequence of subsets in a set Ω. We have

$$\liminf_{i \to \infty} E_i = \{\omega \in \Omega : \omega \text{ belongs to } E_i \text{ for all but finitely many } i\},$$

$$\limsup_{i \to \infty} E_i = \{\omega \in \Omega : \omega \text{ belongs to } E_i \text{ infinitely often}\}.$$

Proof We note that the condition of an outcome ω being in E_i for all but finite many i is equivalent to the statement that ω belongs to E_i eventually for all $i \geq N$, for some integer N. Using the definition of liminf, we see that an outcome ω belongs to $\bigcup_{j \geq 1} \bigcap_{k \geq j} E_k$ if and only if it belongs to $\bigcap_{k \geq j} E_k$ for some j. This is the same as saying that ω belongs to E_k eventually for all $k \geq j$.

To see that an element in $\limsup_i E_i$ occurs in E_i infinitely often, we define $F_j \triangleq \bigcup_{k \geq j} E_k$ and write $\limsup_{i \to \infty} E_i \triangleq \bigcap_{j=1}^{\infty} F_j$.

Consider an element ω in the limsup of E_i's. It must be in F_j for all j. In particular, it must be in F_1. Since F_1 is the union of all E_k's, we can find an index k_1 such that ω in E_{k_1}. Next, we use the fact that ω is in F_{k_1+1}. Since F_{k_1+1} is the union of all E_i's for $i \geq k_1 + 1$, we can find an index $k_2 > k_1$ such that $\omega \in E_{k_2}$. This process can be repeated indefinitely. Therefore, ω belongs to E_{k_1}, E_{k_2}, E_{k_3}, ..., for some strictly increasing indices $k_1 < k_2 < k_3 < \cdots$. This shows that ω occurs in E_i infinitely often, as desired.

Conversely, if ω belongs to E_i for infinitely many indices i, then ω belongs to F_j for all indices j. This is because F_j contains all E_i's for $i \geq j$, and since ω belongs to infinitely many E_i's, it must belong to F_j for all j. Therefore, ω is in $\bigcap_{j \geq 1} F_j$, which is the limsup of E_i's. ∎

2.3   Measure Functions

We will show in the last section of this chapter that it can lead to logical contradictions if we attempt to define a probability measure for all subsets in a sample space. To avoid this, probability theory adopts the solution of defining the probability only for some selected subsets.

The σ-algebra is significant because it serves as the domain of a measure function in probability theory. The measure function is a set function that takes a set as input and returns a nonnegative real number as output, satisfying some axioms.

Definition 2.5

Let $\mathscr{F}$ be a σ-field on a sample space Ω. A set function m from $\mathscr{F}$ to $[0, \infty]$ is called a *measure* if it satisfies the following properties:

1. $m(\emptyset) = 0$.
2. For any sequence of mutually disjoint $A_i \in \mathscr{F}$, for $i = 1, 2, 3, \ldots$, we have

$$m\left(\biguplus_{i=1}^{\infty} A_i\right) = \sum_{i=1}^{\infty} m(A_i).$$

This property is known as the σ-*additive property*.

A measure m is said to be *finite* if $m(\Omega)$ is finite. If $m(\Omega) = 1$, then m is called a *probability measure*.

A *measure space* is a triple $(\Omega, \mathscr{F}, m)$ where $\mathscr{F}$ is a σ-field on Ω and m is a measure on $\mathscr{F}$. A *probability space* is a measure space $(\Omega, \mathscr{F}, m)$ when $\mathscr{F}$ is a σ-field on Ω and m is a probability measure.

Convention with ∞

In Definition 2.5, the measure function m may take the value ∞. We adopt the usual convention for arithmetic operations involving of ∞, namely:

- $c + \infty = \infty$ for all real numbers c.
- $\infty + \infty = \infty$.

Example 2.3.1 (Counting Measure)

Let Ω be any countable set, and let the power set of Ω, $\mathscr{P}(\Omega)$, be the σ-field $\mathscr{F}$. Define the *counting measure* μ of a set $A \subseteq \Omega$ to be the cardinality of A. Then $(\Omega, \mathscr{P}(\Omega), \mu)$ is a measure space.

Example 2.3.2 (Discrete Probability Space)

Let $\Omega = \mathbb{N}$. Suppose $p_1, p_2, p_3, \ldots$ is a sequence of nonnegative real numbers such that $\sum_{i=1}^{\infty} p_i = 1$. For any $E \subseteq \Omega$, let $P(E) = \sum_{i \in E} p_i$. Then $(\Omega, \mathscr{P}(\Omega), P)$ is a probability space.

The first two properties below are elementary, and they hold more generally for any pre-measure (which will be defined in the next chapter).

Theorem 2.2 (Monotonicity)

Let $(\Omega, \mathscr{F}, \mu)$ be a measure space, and suppose A and B are $\mathscr{F}$-measurable sets such that $A \subseteq B$. Then $\mu(A) \leq \mu(B)$.

Proof Since A and $B \setminus A$ are disjoint and $B = A \cup (B \setminus A)$, we have

$$\mu(B) = \mu(A) + \mu(B \setminus A).$$

Since $\mu(B \setminus A)$ is nonnegative, it follows that $\mu(B) \geq \mu(A)$. ∎

Theorem 2.3 (Finite Subadditivity)

Let $(\Omega, \mathscr{F}, \mu)$ be a measure space, and suppose A and B are $\mathscr{F}$-measurable sets. Then

$$\mu(A \cup B) \leq \mu(A) + \mu(B).$$

Proof $\mu(A \cup B) = \mu(A \uplus (B \setminus A)) = \mu(A) + \mu(B \setminus A) \leq \mu(A) + \mu(B)$. ∎

The next general property is about an increasing and decreasing sequence of events.

Definition 2.6

A sequence of sets A_i, for $i = 1, 2, 3, \ldots$, is said to be *increasing* if $A_1 \subseteq A_2 \subseteq A_3 \subseteq \cdots$, or *decreasing* if $A_1 \supseteq A_2 \supseteq A_3 \supseteq \cdots$.

Theorem 2.4 (Lower Semi-Continuity)
Suppose $A_1 \subseteq A_2 \subseteq A_3 \subseteq \cdots$ is a sequence of increasing sets in $\mathscr{F}$ and $(\Omega, \mathscr{F}, \mu)$ is a measure space. Then

$$\lim_{k \to \infty} \mu(A_k) = \mu\left(\bigcup_{i=1}^{\infty} A_i\right).$$

Proof Let $B_1 = A_1$ and $B_i = A_i \setminus A_{i-1}$ for $i \geq 2$. The sets B_i's are mutually disjoint by construction. Moreover, we have $\biguplus_{i=1}^{\infty} B_i = \bigcup_{i=1}^{\infty} A_i$. This gives

$$\lim_{k \to \infty} \mu(A_k) = \lim_{k \to \infty} \mu(B_1 \uplus B_2 \uplus \cdots \uplus B_k)$$

$$= \lim_{k \to \infty} \sum_{i=1}^{k} \mu(B_i)$$

$$\triangleq \sum_{i=1}^{\infty} \mu(B_i) = \mu\left(\biguplus_{i=1}^{\infty} B_i\right) = \mu\left(\bigcup_{i=1}^{\infty} A_i\right),$$

where the second last equality follows from the countable additivity of measure μ. ∎

Theorem 2.5 (Upper Semi-Continuity)
Suppose $A_1 \supseteq A_2 \supseteq A_3 \supseteq \cdots$ is a sequence of decreasing sets in $\mathscr{F}$, and $(\Omega, \mathscr{F}, \mu)$ is a measure space. Furthermore, suppose $\mu(A_1) < \infty$. Then

$$\lim_{k \to \infty} \mu(A_k) = \mu\left(\bigcap_{i=1}^{\infty} A_i\right).$$

Proof Apply Theorem 2.4 to the sequence of events $E_i \triangleq A_1 \setminus A_i$, for $i = 1, 2, 3, \ldots$, and exploit that fact that $\mu(E_i) = \mu(A_1) - \mu(A_i)$. ∎

The properties in Theorems 2.4 and 2.5 are also known as *continuity from below* and *continuity from above*, respectively.

It is a customary to use the notation $A_k \nearrow A$ to signify that $(A_k)_{k=1}^{\infty}$ is a sequence of increasing sets with union A. Similarly, we write $A_k \searrow A$ for a sequence of decreasing sets with intersection A.

2.4　　Borel Sets

In this section, we demonstrate how to construct a σ-field that contains a given collection of subsets in Ω. In particular, suppose we are interested in a topological space. The open sets in the topological space may not satisfy all the requirements of a σ-algebra. In this case, we define the *Borel algebra* as the smallest σ-field that contains all open sets in the given topological space.

> **Theorem 2.6**
>
> Let $\mathscr{C}$ be a collection of subsets in Ω, and $\mathscr{F}$ be the intersection of all σ-fields that contain $\mathscr{C}$ as a sub-collection. Then $\mathscr{F}$ is a σ-field, and it is the smallest σ-field that contains $\mathscr{C}$.

Proof We can write $\mathscr{F}$ as

$$\mathscr{F} = \bigcap_{\substack{\mathscr{G}:\sigma\text{-field} \\ \mathscr{C} \subseteq \mathscr{G}}} \mathscr{G},$$

where the intersection taken over all σ-fields that contain $\mathscr{C}$ as a sub-collection. The intersection is not empty because $\mathscr{P}(\Omega)$ is such a σ-field. We next check that $\mathscr{F}$ is a σ-field.

To show that $\mathscr{F}$ is a σ-field, we need to verify the three axioms:

- $\Omega \in \mathscr{F}$ because Ω is an element of every σ-field, in particular, every σ-field that contains $\mathscr{C}$.
- If $A \in \mathscr{F}$, then A is contained in every σ-field that contains $\mathscr{C}$. Since any σ-field is closed under taking complements, A^c is also contained in every σ-field that contains $\mathscr{C}$. This proves that $A^c \in \mathscr{F}$.
- Suppose $A_i \in \mathscr{F}$, for $i = 1, 2, \ldots$. By definition, A_i is contained in every σ-field that contains $\mathscr{C}$, so their union $\cup_i A_i$ is also contained in every σ-field containing $\mathscr{C}$. Therefore $\cup_i A_i$ is an element of $\mathscr{F}$.

This proves that $\mathscr{F}$ is a σ-field that contains $\mathscr{C}$. Moreover, if $\mathscr{G}$ is any σ-field that contains $\mathscr{C}$, then $\mathscr{G}$ is one of the σ-fields being intersected to form $\mathscr{F}$. Therefore, $\mathscr{F}$ is a subset of $\mathscr{G}$. This implies that $\mathscr{F}$ is the smallest σ-field containing $\mathscr{C}$. ∎

> **Definition 2.7**
>
> The σ-field $\mathscr{F}$ in Theorem 2.6 is called the *σ-field generated* by $\mathscr{C}$ and is denoted by $\sigma(\mathscr{C})$. If $(B_i)_{i \geq 1}$ is a sequence of sets in Ω, then the σ-field generated by this sequence of sets is denoted by $\sigma((B_i)_{i \geq 1})$.

Example 2.4.1 (Generating a σ-algebra in a Finite Measure Space)
Consider a sample space $\Omega = \{a, b, c, d, e, f\}$. Let $A_1 = \{a, b, c\}$ and $A_2 = \{c, d, e\}$ be two of its subsets. We aim to list all the sets in the σ-algebra generated by $\{A_1, A_2\}$. Since the sample space is finite, a σ-algebra is the same as an algebra.

Let $\mathscr{F}$ denote the algebra generated by A_1 and A_2. We will list the subsets of $\mathscr{F}$ in pairs, along with their complement. First, note that $\mathscr{F}$ must contain the empty set, A_1, A_2, and their complements. Therefore, the following sets

$$\emptyset, \{a, b, c, d, e, f\},$$
$$A_1 = \{a, b, c\}, \ A_1^c = \{d, e, f\},$$
$$A_2 = \{c, d, e\}, \ A_2^c = \{a, b, f\},$$

are in $\mathscr{F}$.

Because $\mathscr{F}$ is an algebra and is closed under taking union and intersection, it must contain the union and intersection of any pair of the subsets above and their complements. Therefore, the following sets should also belong to $\mathscr{F}$:

$$\{a, b, c, d, e\}, \{f\}, \qquad\qquad \{c\}, \{a, b, d, e, f\},$$
$$\{a, b, c, f\}, \{d, e\}, \qquad\qquad \{a, b\}, \{c, d, e, f\}.$$

We have listed 14 subsets so far, and we expect that there the number of subsets in $\mathscr{F}$ is a power of 2. Indeed, the last two subsets in $\mathscr{F}$ are

$$\{a, b, d, e\}, \{c, f\}.$$

Note that $\{a, b, d, e\}$ is the symmetric difference of A_1 and A_2. One can check that these subsets are closed under union, intersection, and complement and hence form an algebra. This is the smallest algebra that contains A_1 and A_2.

Definition 2.8

Let $\Omega = \mathbb{R}$ and $\mathscr{C}$ be the collection of all open intervals (a, b), with $a < b$. The σ-field generated by $\mathscr{C}$ is a σ-field called the *Borel field*, or the *Borel algebra*, and is written as $\mathscr{B}(\mathbb{R})$. Likewise, for integer $d \geq 1$, we define the Borel algebra $\mathscr{B}(\mathbb{R}^d)$ as the σ-algebra generated by the open balls in $\mathbb{R}^d$. A set in a Borel algebra is called a *Borel set*.

The Borel algebra $\mathscr{B}(\mathbb{R})$ contains all subsets that we need in practice. For example, closed intervals $[a, b]$ are in $\mathscr{B}(\mathbb{R})$ because it can be expressed as an intersection

$$[a, b] = \bigcap_{n=1}^{\infty} (a - \frac{1}{n}, b + \frac{1}{n}).$$

Semi-infinite intervals $(-\infty, b]$ are in $\mathscr{B}(\mathbb{R})$ because

$$(\infty, b] = \bigcup_{n=1}^{\infty} [b - n, b].$$

Similarly, $[a, \infty)$, (a, ∞), $(-\infty, b)$, $(a, b]$, $[a, b)$ are Borel sets.

In Definition 2.8, we generate the Borel sets by open intervals. Nevertheless, there is nothing special about open intervals, and we can use other types of intervals instead.

Theorem 2.7

The Borel algebra on $\mathbb{R}$ can be generated by closed intervals $[a, b]$, for $a < b$.

Proof Let $\mathscr{C}$ denote the collection of open intervals, and let $\mathscr{C}'$ be the collection of closed intervals in $\mathbb{R}$. We want to show that $\sigma(\mathscr{C}) = \sigma(\mathscr{C}')$. This is equivalent to prove (i) $\sigma(\mathscr{C}) \subseteq \sigma(\mathscr{C}')$ and (ii) $\sigma(\mathscr{C}) \supseteq \sigma(\mathscr{C}')$.

By the definition of $\sigma(\mathscr{C}')$, we have

$$\sigma(\mathscr{C}') = \bigcap_{\substack{\mathscr{H}:\sigma\text{-field} \\ \mathscr{C}' \subseteq \mathscr{H}}} \mathscr{H}.$$

As we have shown in the paragraph before the theorem, the σ-algebra $\sigma(\mathscr{C})$ contains all closed intervals and is one of the σ-algebras $\mathscr{H}$ in the intersection. This proves that $\sigma(\mathscr{C}') \subseteq \sigma(\mathscr{C})$.

In the other direction, the σ-algebra $\sigma(\mathscr{C}')$ contains all open intervals because an open interval (a, b) can be written as

$$(a, b) = \bigcup_{n=1}^{\infty} [a + \frac{1}{n}, b - \frac{1}{n}].$$

Therefore $\sigma(\mathscr{C}) \subseteq \sigma(\mathscr{C}')$. ∎

Similarly, one can show that $\mathscr{B}(\mathbb{R})$ can be generated by sets in the form $(a, b]$, or sets in the form $(-\infty, b]$, etc.

As in the one-dimensional case, $\mathscr{B}(\mathbb{R}^n)$ can be generated by other classes of sets, such as closed n-dimensional boxes, or open n-dimensional balls.

> **Theorem 2.8**
> *For positive integer n, the Borel algebra on $\mathbb{R}^d$ is generated by open d-dimensional open boxes in the form*
>
> $$(a_1, b_1) \times (a_2, b_2) \times \cdots \times (a_d, b_d).$$

The proof is similar to the proof of Theorem 2.7. We just need to show that an open disc in $\mathbb{R}^d$ can be expressed as a countable union of open d-dimensional boxes and *vice versa*.

2.5 Vitali Set

In this section, we prove the existence of a non-measurable set in the interval $[0, 1]$, with respect to the usual length function. To do so, we define the "modulo-1 addition" operation, denoted by $x+y$ mod 1, which takes the fractional part of $x+y$, ensuring that the results are always between 0 and 1. Geometrically, we identify the two endpoints of $[0, 1]$ and regard the interval as a circle, where the mod-1 addition corresponds to a rotation.

We also define the *translation* of a set A by a distance d as

$$A + d \triangleq \{x + d \text{ mod } 1 : x \in A\}.$$

We will assume a technical property, namely that the measure function is translation-invariant (or rotation-invariant, if we are picturing the space as a circle). A set function m is said to be *translation-invariant* if $m(A) = m(A + d)$ holds for any set A in the domain of m and any real number d. We will assume that a set A and any translation $A + d$ of A have the same measure. This property is natural to expect because the usual arc length function on the circle is rotational-invariant.

The following theorem shows that it is impossible to define a probability measure that models the uniform distribution for all subsets of $[0, 1]$. The set V constructed in the proof of the theorem is known as the Vitali set.

> **Theorem 2.9**
> *There is no set function $m : \mathscr{P}([0, 1]) \to [0, \infty)$ that satisfies:*
>
> *1. Monotonicity*
> *2. Countable additivity*

(continued)

Theorem 2.9 (continued)
3. *Translation-invariance*
4. $m([a, b]) = b - a$ *for all* $a < b$

Proof Because 0 and 1 are regarded as the same modulo 1, we will consider the interval $[0, 1)$ in the following discussion, so that all points in $[0, 1)$ are distinct modulo 1. Since we can include a single point in an arbitrarily small closed interval, by the assumptions of monotonicity, a single point has measure zero and hence is negligible.

Define a relation on $[0, 1)$ by $x \sim y \Leftrightarrow x - y \in \mathbb{Q}$. It can be shown that this is an equivalence relation, satisfying the following properties:

(i) (reflexive) $x \sim x$;
(ii) (symmetric) If $x \sim y$, then $y \sim x$;
(iii) (transitive) If $x \sim y$ and $y \sim z$, then $x \sim z$.

We denote the equivalence class that contains x by $[x] \triangleq \{z \in [0, 1) : z \sim x\}$. For example, the equivalence class $[0]$ that contains 0 consists of all rational numbers in the interval $[0, 1)$.

We construct a subset of $[0, 1)$ that contains exactly one representative from each equivalence class. We denote the resulting infinite set by V. By the Axiom of Choice, we can always choose a representative for each class, even though there are uncountably many equivalence classes. We choose 0 to be the representative for the equivalence class $[0] = \mathbb{Q} \cap [0, 1)$. Therefore, the elements of V are all irrational, except for 0.

We claim that the translations $V + r$ of V, with r ranging over all rational numbers in $[0, 1)$, form a partition of $[0, 1)$, i.e.,

$$[0, 1) = \biguplus_{r \in \mathbb{Q} \cap [0,1)} (V + r). \tag{2.1}$$

We first prove that the translated sets $V + r_1$ and $V + r_2$ are disjoint for any two distinct rational numbers r_1 and r_2 in $[0, 1)$. Suppose for the sake of contradiction that α is a common element in $V + r_1$ and $V + r_2$. Then there exist v_1 and v_2 in V such that $\alpha = v_1 + r_1 = v_2 + r_2 \mod 1$. This implies that $v_1 = v_2 + (r_2 - r_1) \mod 1$. Since $r_2 - r_1 \neq 0 \mod 1$, this contradicts the assumption that v_1 and v_2 are the representatives of two different equivalence classes. Therefore, the sets $V + r_1$ and $V + r_2$ are disjoint.

Next, we show that any real number x in $[0, 1)$ belongs to $V + r$ for some rational number $r \in [0, 1)$. Let y be the representative of the coset that contains x. We distinguish two cases. If $x \geq y$, then the difference $r_0 \triangleq x - y$ is a rational number in $[0, 1)$, and $x \in V + r_0$ since x can be written as $x = y + (x - y) = y + r_0$

mod 1. If $x < y$, then $r_1 \triangleq 1 - y + x$ is a rational number in $[0, 1)$, and we have $x \in V + r_1$, as we can write x as $x = y + r_1 - 1$. This completes the proof of the claim in (2.1).

Since the rational numbers in $[0, 1)$ are countable, the union in (2.1) is a countable and disjoint union. By the assumption of countable additivity and translation-invariance, we obtain

$$m([0, 1)) = m\left(\biguplus_{r \in \mathbb{Q} \cap [0,1)} (V + r) \right) = \sum_{r \in \mathbb{Q} \cap [0,1)} m(V). \tag{2.2}$$

If m is defined for all sets, then the value of $m(V)$ is well-defined and equal to a constant c. If $c = 0$, we obtain from (2.2) that $m([0, 1)) = 0$, which contradicts the fact that $m([0, 1)) \geq m([0, 1/2]) = 1/2$. On the other hand, if $c > 0$, we get $m([0, 1)) = \infty$ from (2.2), which contradicts $m([0, 1)) \leq m([0, 1]) = 1$. ∎

This theorem says that even with the simplest uniform distribution, there exists a set to which we cannot assign any probability. This means that we have to abandon some of the assumptions in Theorem 2.9. Since monotonicity, countable additivity, translation-invariance are desirable properties, and we certainly want the probability of interval $[a, b]$ to be $b - a$ in the uniform distribution, we cannot require that a measure function is defined for all subsets. Instead, we assign probabilities only to the subsets in a σ-algebra. Subsets that are not in the σ-algebra have no probability.

Problems

2.1. Show that a set function $\mu : \mathscr{F} \to [0, \infty]$ is a measure if and only if:

1. μ is σ-additive.
2. There exists a set $A \in \mathscr{F}$ such that $\mu(A)$ is finite.

2.2.

(a) Let $A_1, A_2, A_3, \ldots$ be a sequence of measurable sets with measure 0 in a measure space. Show that the union $\cup_{i=1}^{\infty} A_i$ has measure 0.
(b) Prove that if $B_1, B_2, B_3, \ldots$ is a sequence of events with probability 1 in a probability space, then the intersection $\cap_{i=1}^{\infty} B_i$ has probability 1.

2.3. We define the Borel algebra $\mathscr{B}(\mathbb{R})$ as the smallest σ-algebra on $\mathbb{R}$ that contains all open intervals. Show that $\mathscr{B}(\mathbb{R})$ can be generated by any one of the following collection of sets:

- $\{(-\infty, b] : b \in \mathbb{R}\}$.
- $\{(-\infty, b) : b \in \mathbb{R}\}$.
- $\{(a, \infty) : a \in \mathbb{R}\}$.

- $\{[a, \infty) : a \in \mathbb{R}\}$.
- $\{(a, b] : a, b \in \mathbb{R}\}$.

2.4. (Union bound) For any n events $A_1, A_2, \ldots, A_n$, prove that

$$\Pr(A_1 \cup A_2 \cup \cdots \cup A_n) \le \sum_{i=1}^{n} \Pr(A_i).$$

Use the property of continuity from below to prove that, for any sequence of $\mathscr{F}$-measurable sets A_n, for $n \ge 1$, we have

$$\Pr\left(\bigcup_{i=1}^{\infty} A_i\right) \le \sum_{i=1}^{\infty} \Pr(A_i).$$

2.5. Consider the counting measure μ on a countable sample space Ω. Construct a sequence of decreasing sets $A_1 \supseteq A_2 \supseteq A_3 \supseteq \cdots$ such that $\bigcap_{i=1}^{\infty} A_i = \emptyset$, but $\lim_{i \to \infty} \mu(A_i) \ne 0$.

2.6. For $i = 1, 2, 3, \ldots$, let A_i be a subset of a set Ω. Prove that

$$\liminf_{i \to \infty} A_i \subseteq \limsup_{i \to \infty} A_i.$$

2.7. Find the limits superior and inferior of the following sequences of sets $(A_i)_{i \ge 1}$:

(a) $A_i = [1/i, 4 + (-1)^i]$.
(b) A_i is the interval $[0, 1]$ if i is odd, and $[1, 2]$ if i is even.

2.8. Consider a sample space $\Omega = \{a, b, c, d, e, f\}$ consisting of 6 elements. Let A_1 and A_2 denote the subsets $\{a, c, d\}$ and $\{c, d, f\}$, respectively. List the subsets in the σ-algebra generated by A_1 and A_2.

2.9. Let $A_1, A_2, \ldots, A_n$ be n arbitrary subsets of Ω. Describe explicitly the σ-field generated by $A_1, A_2, \ldots, A_n$, and provide an upper bound for the number of subsets it contains.

2.10. Show that the Borel algebra on $\mathbb{R}$ can be generated by countably many sets in $\mathbb{R}$.

2.11. Verify that the Cantor set is a Borel set.

2.12. Suppose $\mathscr{A}$ is a collection of subsets in Ω such that (a) $\Omega \in \mathscr{A}$, (b) $\mathscr{A}$ is closed under taking set difference, i.e., $A, B \in \mathscr{A}$ implies $B \setminus A \in \mathscr{A}$. Show that $\mathscr{A}$ is a field.

Lebesgue–Stieltjes Measures

3

Defining a measure function on the Borel algebra is a non-trivial task. It is impractical to explicitly specify the measure of each Borel set due to the lack of an explicit description of all Borel sets. Instead, we can begin with an algebra of sets that is closed under finite set operations, such as union and intersection. By using a Stieltjes measure function, we can define a pre-measure on this algebra that satisfies properties analogous to those of a measure function, such as monotonicity and additivity. Then, by applying the measure extension theorem, we can obtain a measure that is defined on the Borel algebra. This provides a method for constructing Borel measure on the real number line, encompassing all types of distributions, including that are discrete, continuous, or a mixture of both, as well as singular distributions.

3.1 Pre-measure

A pre-measure is a set function that takes a set as input and returns a nonnegative real number. Unlike a measure function, the domain of a pre-measure is an algebra or a field, rather than a σ-algebra.

Definition 3.1

A *pre-measure* μ_0 defined on a field $\mathscr{F}_0$ is a set function that maps $\mathscr{F}_0$ to $[0, \infty]$, satisfying the following conditions:

1. $\mu_0(\emptyset) = 0$.
2. If $A_i \in \mathscr{F}_0$ for $i = 1, 2, 3, \ldots$ are mutually disjoint sets in Ω and $\biguplus_i A_i \in \mathscr{F}_0$, then

© The Author(s), under exclusive license to Springer Nature Switzerland AG 2023
K. Shum, *Measure-Theoretic Probability*, Compact Textbooks in Mathematics,
https://doi.org/10.1007/978-3-031-49830-5_3

$$\mu_0\left(\biguplus_{i=1}^{\infty} A_i\right) = \sum_{i=1}^{\infty} \mu_0(A_i).$$

▶ **Remark** The conditions in Definition 3.1 ensure that the pre-measure satisfies some of the basic properties of a measure, such as nonnegativity and countable additivity. However, since $\mathscr{F}_0$ is only assumed to be a field, the union $\uplus_i A_i$ may or may not be in $\mathscr{F}_0$. For this reason, the condition $\uplus_i A_i \in \mathscr{F}_0$ is included in the definition of pre-measure to ensure that the pre-measure of the union is well-defined. If $\uplus_i A_i$ is not in $\mathscr{F}_0$, then the condition is vacuously true.

Example 3.1.1 (Example of a Pre-measure that is Not a Measure)
Recall that one way to define a field on $\Omega = \mathbb{N}$ is to define it as the collection of all finite and co-finite sets. We can define a pre-measure on this field by setting $\mu_0(E) = \infty$ if E is infinite, and $\mu_0(E)$ to be the cardinality of E if E is finite.

Example 3.1.2 (A Pre-measure on $\mathbb{R}$)
Let $\Omega = \mathbb{R}$, and consider the collection of sets $\mathscr{B}_0$ in Ω that can be expressed as a finite disjoint union of intervals of the form

$$(a_1, b_1] \uplus (a_2, b_2] \uplus \cdots \uplus (a_n, b_n], \tag{3.1}$$

where n is a positive integer, and a_1 or b_n may be infinite. One can check that this collection of sets is an algebra, and it is closed under complement and finite union. For example, the complement of $(a, b]$ in $\mathbb{R}$ is the union of $(-\infty, a]$ and $(b, \infty]$.

If $a_1 = -\infty$ or $b_n = \infty$, we define the length of the set in (3.1) to be ∞. Otherwise, if all a_i and b_i in (3.1) are finite, we define the total length of the set to be

$$(b_1 - a_1) + (b_2 - a_2) + \cdots + (b_n - a_n).$$

This length function is a pre-measure on $\mathscr{B}_0$.

Definition 3.2

Suppose μ_0 is a pre-measure defined on a field $\mathscr{F}_0$, and let $\mathscr{F}$ be the σ-field generated by $\mathscr{F}_0$. We say that $\mu : \mathscr{F} \to [0, \infty]$ is an *extension* of μ_0 if μ is a measure and satisfies

$$\mu(E) = \mu_0(E), \quad \text{for all } E \in \mathscr{F}_0.$$

In other words, the extension of μ_0 to $\mathscr{F}$ agrees with μ_0 on the field $\mathscr{F}_0$.

In the measure extension theorem, we require a technical condition on the pre-measure called σ-finiteness. A measure function is regarded as well-behaved in general if it is σ-finite. We note that a probability measure is automatically σ-finite.

Definition 3.3

A pre-measure μ_0 defined on $\mathcal{F}_0$ is said to be σ-*finite* if we can find at most countably many sets $\Omega_i \in \mathcal{F}_0$, for $i = 1, 2, 3, \ldots$, such that $\cup_i \Omega_i = \Omega$, and $\mu_0(\Omega_i) < \infty$ for all i.

Without loss of generality, we may assume that the sets Ω_i in the definition of σ-finiteness form a partition of Ω. If the Ω_i's are not mutually disjoint, we can re-define Ω_i by setting

$$\Omega_i := \Omega_i \setminus (\cup_{j=1}^{i-1} \Omega_j).$$

The following theorem is the main theorem in this chapter.

Theorem 3.1 (Hahn–Kolmogorov–Carathéodory)
Let $\mathcal{F}_0$ be a field on a sample space Ω. Given a pre-measure μ_0 defined on $\mathcal{F}_0$, there is a measure μ that extends to $\sigma(\mathcal{F}_0)$.
  Moreover, if μ_0 is σ-finite, then the extension is unique, i.e., if μ and μ' are two extensions of μ_0, we have $\mu(E) = \mu'(E)$ for all $E \in \mathcal{F}$.

The proof of the uniqueness of the extension in Theorem 3.1 will be given in a subsequent section. The proof of the existence can be found in standard textbooks on measure theory, such as [2, Theorem 11.2] and [4, Theorem 1.1.9].

Example 3.1.3 (Counting Measure)
The pre-measure in Example 3.1.1 is σ-finite, because $\mathbb{N}$ can be covered by countably many singletons $\{i\}$, with i running over all positive integers. We can take $\Omega_i = \{i\}$ for $i = 1, 2, 3, \ldots$. This pre-measure can be extended to the power set $2^{\mathbb{N}}$, and the resulting measure is the counting measure on $\mathbb{N}$.

Example 3.1.4 (Borel Measure)
The pre-measure in Example 3.1.2 is σ-finite. We can take $\Omega_i \subset \mathbb{R}$ to be the interval $(-i, i]$ in Definition 3.3, which covers $\mathbb{R}$ and has finite measure. By Theorem 3.1, the pre-measure can be extended to the Borel algebra $\mathcal{B}(\mathbb{R})$, and the extension is unique. This measure is called the *Borel measure* on $\mathbb{R}$ and is denoted by λ.

3.2 Stieltjes Measure Function

We begin by considering a probability space $(\mathbb{R}, \mathcal{B}(\mathbb{R}), P)$ defined on the real number line $\mathbb{R}$, equipped with the Borel algebra $\mathcal{B}(\mathbb{R})$. The distribution function provides a convenient way to visualize a probability measure.

Definition 3.4

Let $(\mathbb{R}, \mathscr{B}(\mathbb{R}), P)$ be a probability space. The *distribution function* induced by P is defined as

$$F(x) \triangleq P((-\infty, x]) \qquad \text{for } x \in \mathbb{R}.$$

▶ **Distribution Function Vs. Cumulative Distribution Function** The function $F(x)$ in Definition 3.4 is very similar to the cumulative distribution function (cdf) of a random variable. The main difference is that a cdf is defined for a random variable, whereas a distribution function is defined for a probability measure.

Convention

Some people prefer to define the distribution function using open intervals of the form $(-\infty, x)$, instead of the left-open-right-closed intervals $(-\infty, x]$. In this case, the distribution function is defined as $P((-\infty, x))$ for $x \in \mathbb{R}$. The choice between using one type of intervals or the other is a matter of convention.

Theorem 3.2

The distribution function $F(x)$ of a probability measure satisfies the following properties:

1. *$F(x)$ is non-decreasing.*
2. *$F(x)$ is continuous from the right.*
3. *$\lim_{x \to \infty} F(x) = 1$.*
4. *$\lim_{x \to -\infty} F(x) = 0$.*

Proof Property 1 follows from the monotonic property of measure. Property 2 is a consequence of the upper semi-continuity. If $(x_i)_{i=1}^{\infty}$ is a decreasing sequence converging to x from the right, then $(-\infty, x_i]$ is a decreasing sequence of sets, and the intersection is $(-\infty, x_i]$. Hence, $\lim_{i \to \infty} F(x_i) = \lim_{i \to \infty} P((-\infty, x_i]) = P((-\infty, x]) = F(x)$. Properties 3 and 4 also follow from semi-continuity properties of probability measure.   ∎

Motivated by the previous theorem, we define a Stieltjes measure function in general.

Definition 3.5

A function $F : \mathbb{R} \to \mathbb{R}$ is called a *Stieltjes measure function* if:

1. F is non-decreasing.
2. F is continuous from the right, i.e., $\lim_{y \to x+} F(y) = F(x)$ for all $x \in \mathbb{R}$.

A Stieltjes measure function does not necessarily have a density function on the real number line. However, if F is both continuous and differentiable, then the derivative $f(x) = F'(x)$ is a density function, and an interval $[a, b]$ in the real number line has measure $\int_a^b f(x)\,dx$.

In general, as we will show in the next section, a Stieltjes measure function defines a measure on the Borel algebra of $\mathbb{R}$, called the Lebesgue–Stieltjes measure, which assigns to each left-open-right-closed interval $(a, b]$ the value $F(b) - F(a)$.

Example 3.2.1 (Examples of Stieltjes Measure Functions)

1. The identity function $F(x) = x$ is a Stieltjes measure function, since it is both continuous and non-decreasing.
2. Fix a constant x_0. The function

$$F(x) = \begin{cases} 1 & \text{if } x \geq x_0 \\ 0 & \text{if } x < x_0 \end{cases}$$

defines a Stieltjes measure function. The function $F(x)$ is continuous everywhere except at $x = x_0$, where it has a discontinuity jump.
3. The cumulative distribution function $F(x)$ of the standard normal distribution

$$F(x) = \int_{-\infty}^x \frac{1}{\sqrt{2\pi}} e^{-t^2/2}\,dt$$

is a Stieltjes measure function. In fact, if $f(x)$ is the pdf of any continuous-type distribution, the function $F(x) = \int_{-\infty}^x f(t)\,dt$ is a Stieltjes measure function.

3.3 Lebesgue–Stieltjes Measures

Given a Stieltjes measure function, we can apply the measure extension theorem to construct a measure on the Borel algebra $\mathscr{B}(\mathbb{R})$. Intuitively, the Lebesgue–Stieltjes measure μ assigns a length to each Borel set on the real line, where the length of an interval $(a, b]$ is given by the difference in the values of F at the endpoints a and b.

> **Theorem 3.3**
> *Let F be a Stieltjes measure function defined on $\mathbb{R}$. Then there exists a unique measure μ defined on Borel σ-algebra $\mathscr{B}(\mathbb{R})$ such that $\mu((a, b]) = F(b) - F(a)$. The measure μ is called the* Lebesgue–Stieltjes measure *induced by F.*

This theorem is an application of the measure extension theorem (Theorem 3.1). The proof relies on a standard result from real analysis, which we first recall below.

Definition 3.6

A closed and bounded interval in $\mathbb{R}$ is called a *compact* set in $\mathbb{R}$.

Definition 3.7

An *open cover* of a set $A \subseteq \mathbb{R}$ is a collection of open sets $\{A_\alpha\}_{\alpha \in I}$, indexed by an arbitrary set I, such that A is contained in the union of A_α over all $\alpha \in I$. A *subcover* of A is a sub-collection of $\{A_\alpha\}_{\alpha \in I}$ that is also a cover of A. If a cover consists of finitely many open sets, it is said to be a *finite cover*.

To illustrate these definitions, consider the closed interval $[0, 1]$, which is a compact set. Let $\epsilon > 0$ be a fixed real number. We can cover $[0, 1]$ by infinitely many open intervals of the form $(a - \epsilon, a + \epsilon)$, where a ranges over all real numbers in $[0, 1]$. However, there is a lot of redundancy in this open cover, and we can in fact cover $[0, 1]$ using only finitely many open sets from this collection. For instance, we can choose the open intervals $(k\epsilon - \epsilon, k\epsilon + \epsilon)$, where k ranges over $0, 1, 2, \ldots, \lfloor 1/\epsilon \rfloor$.

The Heine–Borel theorem states that the above example is a general phenomenon. We can always find a finite subcover for any compact set.

Theorem 3.4 (Heine–Borel)
Let A be a subset of $\mathbb{R}$. Then A is compact if and only if every open cover of A has a finite subcover.

In the proof of the main result Theorem 3.3 in this section, it is more convenient to show that a set function is σ-additive through σ-subadditivity.

Definition 3.8

(i) A set function μ on a field $\mathscr{F}_0$ is said to be *finitely additive* if for any finite collection of disjoint set $A_1, A_2, \ldots, A_n$ in $\mathscr{F}_0$, we have

$$\mu\left(\biguplus_{i=1}^{n} A_i\right) = \sum_{i=1}^{n} \mu(A_i).$$

(ii) A set function μ on a field $\mathscr{F}_0$ is called *σ-subadditive* if for any countable collection $(A_i)_{i \geq 1}$ of sets in $\mathscr{F}_0$ such that $\cup_{i=1}^{\infty} A_i \in \mathscr{F}_0$, we have

$$\mu\left(\bigcup_{i=1}^{\infty} A_i\right) \leq \sum_{i=1}^{\infty} \mu(A_i).$$

One can show that any σ-additive function satisfies the two conditions in the above definition.

Theorem 3.5
A set function μ defined on a field $\mathscr{F}_0$, with $\mu(\emptyset) = 0$, is σ-additive if:

(i) μ is finitely additive, and
(ii) μ is σ-subadditive.

Hence, a nonnegative set function on a field $\mathscr{F}_0$ that satisfies $\mu(\emptyset) = 0$, finite additivity, and σ-subaddtivity is a pre-measure on $\mathscr{F}_0$.

Proof Suppose $\mu : \mathscr{F}_0 \to [0, \infty]$ is a set function that satisfies finite additivity, σ-subadditivity, and $\mu(\emptyset) = 0$. Suppose E_i are mutually disjoint sets in $\mathscr{F}_0$, for $i = 1, 2, 3, \ldots$, such that $\biguplus_{i=1}^{\infty} E_i = E \in \mathscr{F}_0$. For each finite n, we have

$$\mu(E) \geq \mu(E_1 \uplus E_2 \uplus \cdots \uplus E_n) = \sum_{i=1}^{n} \mu(E_i).$$

Taking the limit as $n \to \infty$, we obtain $\mu(E) \geq \sum_{i=1}^{\infty} \mu(E_i)$.

To show the reverse inequality, we note that μ is σ-subadditivity, and so

$$\mu(E) = \mu\left(\bigcup_{i=1}^{\infty} E_i\right) \leq \sum_{i=1}^{\infty} \mu(E_i).$$

This shows that μ is σ-additive and hence is a pre-measure. ∎

Proof of Theorem 3.3 Suppose F is a Stieltjes measure function and $\mathscr{B}_0$ is the collection of sets as defined in Example 3.1.2, which are disjoint unions of intervals in the form $(a, b]$. We define a set function on $\mathscr{B}_0$ as follows: for any $(a, b]$ in $\mathscr{B}_0$, we set $\mu_0((a, b]) \triangleq F(b) - F(a)$, and we define $\mu_0(\emptyset) = 0$. For any disjoint union $A = \biguplus_{i=1}^{m} (a_i, b_i]$, we set

$$\mu_0(A) \triangleq \sum_{i=1}^{m} [F(b_i) - F(a_i)].$$

Step 1. Show that μ_0 is a well-defined set function.

Suppose a set $A \in \mathscr{B}_0$ can be written in two ways:

$$A = \uplus_{i=1}^m (a_i, b_i] = \uplus_{j=1}^n (c_j, d_j].$$

We want to check that the value of μ_0 on A is independent of the choice of representation. The proof idea is to consider a common refinement of the two partitions.

For each pair (i, j), for $1 \leq i \leq m$ and $1 \leq j \leq n$, define the interval

$$I_{ij} = (a_i, b_i] \cap (c_j, d_j] = (\max(a_i, c_j), \min(b_i, d_j)],$$

which may be empty. It can be verified that the intervals I_{ij} are pair-wise disjoint and their union is equal to A. Furthermore, $\{I_{ij}\}_{i,j}$ is a subdivision of the partitions $\{(a_i, b_i]\}_{i=1}^m$ and $\{(c_j, d_j]\}_{j=1}^n$. Hence,

$$\sum_{i=1}^n \mu_0((a_i, b_i]) = \sum_{i=1}^m \sum_{j=1}^n \mu_0(I_{ij}) = \sum_{j=1}^n \mu_0((c_j, d_j]).$$

This shows that μ_0 is well-defined on $\mathscr{B}_0$.

Step 2. Show that μ_0 is finitely additive.

Since a set in the algebra $\mathscr{B}_0$ is a finite disjoint union of left-open-right-closed intervals, it is sufficient to consider the union of disjoint intervals $(a_i, b_i]$, for $i = 1, 2, \ldots, m$ such that $\uplus_{i=1}^m (a_i, b_i]$ is equal to $(a, b]$. After some re-labeling, we can re-order them and write

$$a = a_1 < b_1 = a_2 < b_2 = a_3 < b_3 \cdots a_n < b_n = b.$$

We have a telescoping sum

$$\sum_{i=1}^n \mu_0((a_i, b_i]) = \sum_{i=1}^n [F(b_i) - F(a_i)] = F(b) - F(a) = \mu_0((a, b]).$$

This proves that μ_0 is a finitely additive.

Step 3. Show that μ_0 is σ-subadditive.

Suppose $(a, b] = \bigcup_{i=1}^\infty (a_i, b_i]$. The subintervals $(a_i, b_i]$'s need not be mutually disjoint. We want to show

$$\mu_0((a, b]) \leq \sum_{i=1}^\infty \mu_0((a_i, b_i]). \tag{3.2}$$

The difficulty in proving (3.2) is that it involves infinitely many intervals. We would like to reduce it to finitely many intervals using the Heine–Borel theorem.

As in a typical proof of real analysis, we fix a positive but arbitrarily small ϵ and make some relaxation. Pick $\delta > 0$ such that $F(a + \delta) < F(a) + \epsilon$. We choose δ small enough so that $a + \delta < b$. Note that the right continuity of F is used when we pick δ.

For each $i = 1, 2, 3, \ldots$, pick $\eta_i > 0$ such that $F(b_i + \eta_i) < F(b_i) + \frac{\epsilon}{2^i}$. We have an open cover

$$[a + \delta, b] \subseteq \bigcup_{i=1}^{\infty} (a_i, b_i + \eta_i).$$

By Heine–Borel theorem, there exist $i_1, i_2, \ldots, i_m$, such that $(a_{i_j}, b_{i_j} + \eta_{i_j})$, for $j = 1, 2, \ldots, m$, form a finite cover of the closed and bounded interval $[a + \delta, b]$,

$$[a + \delta, b] \subseteq \bigcup_{j=1}^{m} (a_{i_j}, b_{i_j} + \eta_{i_j}).$$

We cannot directly apply μ_0 to both sides in the above line because μ_0 is only defined on left-open-right-closed intervals, and their finite unions. To this end, we remove one point on the left side and add m points on the right side to obtain

$$(a + \delta, b] \subseteq \bigcup_{j=1}^{m} (a_{i_j}, b_{i_j} + \eta_{i_j}].$$

We now apply the set function μ_0 to both sides and use the assumption that μ_0 is finitely additive. This gives

$$F(b) - F(a + \delta) \leq \sum_{j=1}^{m} F(b_{i_j} + \eta_{i_j}) - F(a_{i_j}) \leq \sum_{i=1}^{\infty} F(b_i + \eta_i) - F(a_i).$$

In the last step, we have simply added infinitely many nonnegative terms on the right side. We have used the assumption that F is a non-decreasing function in this step.

We express all terms in terms of ϵ and simplify the inequality to

$$F(b) - F(a) - \epsilon \leq \sum_{i=1}^{\infty} \left(F(b_i) - F(a_i) + \frac{\epsilon}{2^i} \right)$$

$$F(b) - F(a) \leq 2\epsilon + \sum_{i=1}^{\infty} \left(F(b_i) - F(a_i) \right).$$

Because ϵ could be arbitrarily small, we obtain

$$F(b) - F(a) \leq \sum_{i=1}^{\infty} \big(F(b_i) - F(a_i) \big).$$

This proves that μ_0 is σ-subadditive.

Step 4. μ_0 is σ-finite.

We can partition the real number line as a union of $(m - 1, m]$, for $m \in \mathbb{Z}$. The pre-measure $\mu_0((m - 1, m])$ is finite for all m.

Combining all the previous steps, we see that μ_0 is a σ-finite pre-measure on the algebra $\mathscr{B}_0$, which generates the Borel algebra on $\mathbb{R}$. By applying the measure extension theorem (Theorem 3.1), we obtain a unique measure on $\mathscr{B}(\mathbb{R})$ that extends μ_0. ∎

Using Theorem 3.3, we can construct the following examples of Lebesgue–Stieltjes measures.

Example 3.3.1 (Lebesgue Measure)
Take $F(x) = x$. The resulting Lebesgue–Stieltjes measure defined on $\mathscr{B}(\mathbb{R})$ is known as the Lebesgue measure. We denote this measure by λ. This measure has the property that $\lambda([a, b]) = b - a$, which can be derived by approaching the closed interval $[a, b]$ by taking intersection of $(a - 1/n, b]$ for $n \geq 1$. Additionally, the Lebesgue measure is translation-invariant.

Example 3.3.2 (Uniform Distribution)
Let $F(x)$ be the function defined by

$$F(x) = \begin{cases} 0 & \text{for } x < 0 \\ x & \text{for } 0 \leq x \leq 1 \\ 1 & \text{for } 1 < x. \end{cases}$$

The resulting Lebesgue–Stieltjes measure is a model for uniform distribution between 0 and 1. For example, for $0 < a < b < 1$, the measure of (a, b) is $b - a$.

Example 3.3.3 (Continuous Distribution)
Let $f(x)$ be a pdf, i.e., a Riemann integrable function with $f(x) \geq 0$ for all x and $\int_{-\infty}^{\infty} f(x)\, dx = 1$. The function $F(x) = \int_{-\infty}^{x} f(t)\, dt$ is a Stieltjes measure function. We denote the corresponding Lebesgue–Stieltjes measure by P, which is defined for all Borel sets in $\mathbb{R}$. In particular, if we pick an open set (a, b), for $a < b$, then the probability measure of (a, b) is equal to $\int_a^b f(t)\, dt$.

Example 3.3.4 (Point Mass (Dirac Measure))
Consider the Stieltjes measure function given by

$$F(x) = \begin{cases} 1 & \text{if } x \geq x_0 \\ 0 & \text{if } x < x_0, \end{cases}$$

where x_0 is a fixed constant. We have a "point mass" at $x = x_0$. Let P denote the resulting probability measure. By writing $\{x_0\}$ as the intersection of $(x_0 - n^{-1}, x_0]$ for $n \geq 1$, we can calculate

$$P(\{x_0\}) = \lim_{n \to \infty} P((x_0 - n^{-1}, x_0]) = \lim_{n \to \infty} F(x_0) - F(x_0 - n^{-1}) = \lim_{n \to \infty} 1 = 1.$$

The measure P has the property that $P(A) = 1$ if $x_0 \in A$ and $P(A) = 0$ if $x_0 \notin A$.

Example 3.3.5 (Discrete Distribution)

For $i = 1, 2, 3, \ldots$, let p_i be positive real numbers satisfying $p_i \geq 0$ and $\sum_{i=1}^{\infty} p_i = 1$. Let x_i be distinct real numbers for $i = 1, 2, 3, \ldots$. We want to create a probability space that has probability p_i on the point x_i. We suppose that the numbers x_i's are discrete, meaning that we can find a sufficiently small radius ϵ such that the intervals $(x_i - \epsilon, x_i + \epsilon)$ are mutually disjoint.

We define a Stieltjes measure function F that is flat except at the points x_i's, and there is a vertical jump of size p_i at x_i, for $i = 1, 2, 3, \ldots$. Let P be the Lebesgue–Stieltjes measure associated with this Stieltjes measure function. Using the same argument as in the last example, we can verify that

$$P(\{x\}) = \begin{cases} p_i & \text{if } x = x_i \text{ for some } i \\ 0 & \text{otherwise.} \end{cases}$$

The set $A = \{x_1, x_2, x_3, \ldots\}$ has P-measure $P(A) = 1$. However, as a countable set, the Lebesgue measure (defined in Example 3.3.1) of this set is 0. The distribution of probability is concentrated on the points x_i's.

Example 3.3.6 (Cantor Distribution)

In Example 1.2.3, we obtain the Cantor distribution from an infinite sequence of independent random variables. In this example, we take the Cantor function (which is also known as the Devil's staircase) as the Stieltjes measure function and apply the measure extension theorem to obtain the Cantor distribution.

The Cantor function can be defined iteratively, starting with

$$F_1(x) = \begin{cases} 3x/2 & \text{if } 0 \leq x \leq 1/3 \\ 0.5 & \text{if } 1/3 < x < 2/3 \\ 0.5 + 0.5(3x - 2) & \text{if } 2/3 \leq x \leq 1. \end{cases}$$

This is a piece-wise linear function with two pieces with slope $1/6$ in the intervals $[0, 1/3]$ and $[2/3, 1]$. The function F_1 is flat in the open interval $I_1 = (1/3, 2/3)$ (See Fig. 3.1).

Given $F_1(x)$, we replace each of the two slant pieces in $F_1(x)$ by a graph with the same shape as $F_1(x)$ and call the resulting function $F_2(x)$. There are three horizontal parts in $F_2(x)$, and they are I_1

$$I_{21} = (1/9, 2/9), \text{ and } I_{22} = (7/9, 8/9).$$

In general, we obtain $F_{i+1}(x)$ by replacing the 2^i non-horizontal parts by a graph that is similar to the graph of $F_1(x)$. After iteration i, we have 2^{i-1} additional intervals of length $1/3^i$ on which the function $F_i(x)$ has slope zero. The Cantor function $F(x)$ is the limit that we obtain by taking $i \to \infty$.

It can be shown that the functions $F_i(x)$ for $i = 1, 2, 3, \ldots$ converge uniformly. Since each $F_i(x)$ is continuous for all i, the limit function $F(x)$ is continuous. Therefore, the Lebesgue–Stieltjes measure P induced by the Cantor function has the property $P(\{x\}) = 0$ for any x. On the other hand, $F(x)$ has zero slope almost everywhere. Indeed, the total length of the intervals $I_1, I_{21}, I_{22}, \ldots$, on which the Cantor function is flat, has Lebesgue measure

Fig. 3.1 The first three
iterations in the construction
of the Cantor function

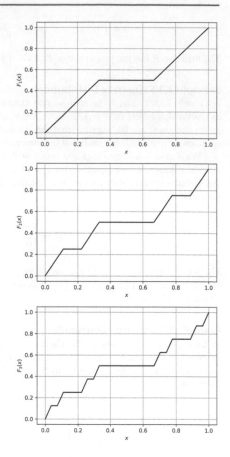

$$\sum_{i=1}^{\infty} 2^{i-1}\frac{1}{3^i} = \frac{1}{2}\sum_{i=1}^{\infty}(2/3)^i = 1.$$

As a result, the complement of the union of intervals $I_1 \cup I_{21} \cup I_{22} \cup \cdots$ in [0, 1], denoted by C, has Lebesgue measure 0. Nonetheless, the probability measure of C is equal to 1. This is the notorious example of singular distribution with continuous distribution function.

In this section, we consider the construction of Lebesgue–Stieltjes measures on the one-dimensional real number line. The idea can be extended to finite-dimensional Euclidean space $\mathbb{R}^d$, where the Stieltjes measure function will be replaced by a function in d variables. For example, in dimension $d = 2$, we can specify a measure by using a function $F(x, y)$ of two variables x and y and define the measure of a rectangle $(a, b] \times (c, d]$ by $F(b, d) - F(a, d) - F(c, b) + F(a, c)$. Readers interested in further details are referred to [1, Section 1.4] and [4, Section 1.1].

3.4   **Null Sets and Complete Measures**

Consider a probability space $(\Omega, \mathcal{F}, P)$. If an event E has probability zero, we might expect that all subsets of E also have probability zero. However, in some cases, not all subsets of a set are measurable, and therefore, we cannot assign any measure to them. To address this issue, we introduce the following definitions.

Definition 3.9

Given a measure space $(\Omega, \mathcal{F}, \mu)$, we define a set $N \subseteq \Omega$ as a *null set* if there exists a set $E \in \mathcal{F}$ such that $N \subseteq E$ and $\mu(E) = 0$. Note that N need not be $\mathcal{F}$-measurable. The measure space $(\Omega, \mathcal{F}, \mu)$ is said to be *complete* if all null sets are indeed $\mathcal{F}$-measurable.

The concept of a null set depends on the choice of measure μ. A set may be a null set under one measure μ but is not a null set under another measure ν.

Using cardinality argument, one can demonstrate the existence of null sets with respect to the Lebesgue measure that are not Borel sets. Let $\mathfrak{c}$ denote the cardinality of $\mathbb{R}$. One can prove that there are $\mathfrak{c}$ Borel subsets of $\mathbb{R}$, while the Cantor set has cardinality $\mathfrak{c}$ and $2^{\mathfrak{c}}$ subsets. Therefore, there must be a subset of the Cantor set that is not measurable with respect to the Lebesgue measure. However, the proof of this result requires more advanced set theory and is beyond the scope of this book.

To address the issue of non-measurable null sets, we can construct a completion of the measure space.

Theorem 3.6

Let $(\Omega, \mathcal{F}, \mu)$ be a measure space and $\mathcal{N}$ be the set of all null sets. We can define a new collection of sets

$$\overline{\mathcal{F}} \triangleq \{A \cup N : A \in \mathcal{F} \text{ and } N \in \mathcal{N}\}.$$

We can extend the measure μ to a measure on $\overline{\mathcal{F}}$, denoted by μ', by

$$\mu'(A \cup N) \triangleq \mu(A),$$

and the extended measure space $(\Omega, \overline{\mathcal{F}}, \mu')$ is complete.

By enlarging the σ-field in this way, we may assume that the measure is complete without loss of generality.

3.5   Uniqueness of Measure Extension

We will use the term *set system* to refer to a collection of subsets in Ω.

Definition 3.10

We define a π-*system* in Ω as a set system $\mathcal{P}$ that is closed under intersection, i.e., it satisfies

$$A \in \mathcal{P} \text{ and } B \in \mathcal{P} \Rightarrow A \cap B \in \mathcal{P}.$$

A λ-*system* in Ω is a set system $\mathcal{L}$ that satisfies the following conditions:

  (i) $\Omega \in \mathcal{L}$.
 (ii) If $A \in \mathcal{L}$, then $\Omega \setminus A \in \mathcal{L}$.
(iii) If $A_i \in \mathcal{L}$ for $i = 1, 2, 3, \ldots$ is a sequence of mutually disjoint sets, then $\cup_i A_i \in \mathcal{L}$.

A λ-system is also known as a Dynkin system, named after the mathematician Eugene Dynkin, who first introduced this concept. The π-λ theorem, also proved by Dynkin, is a fundamental result in measure theory and probability theory.

The theorem says that if a π-system $\mathcal{P}$ is contained in a λ-system $\mathcal{L}$, then we can enlarge the $\mathcal{P}$ to the σ-field generated by $\mathcal{P}$, and this σ-field is also contained in $\mathcal{L}$. Formally, the π-λ theorem is stated as follows.

Theorem 3.7 (π-λ Theorem)
If $\mathcal{L}$ is a λ-system containing a π-system $\mathcal{P}$, then

$$\sigma(\mathcal{P}) \subseteq \mathcal{L}.$$

The proof of π-λ theorem is an exercise (Exercise 3.9).
We can use the π-λ theorem to prove the uniqueness of measure extension.

Theorem 3.8 (Uniqueness of Measure Extension)
Suppose $\mathcal{F}_0$ is a field on a sample space Ω, and μ_1 and μ_2 are two measures on $\sigma(\mathcal{F}_0)$ such that $\mu_1(A) = \mu_2(A)$ for all $A \in \mathcal{F}_0$. Furthermore, suppose there exists a sequence of disjoint sets $\Omega_i \in \mathcal{F}_0$, for $i = 1, 2, 3, \ldots$, such that $\uplus_i \Omega_i = \Omega$ and $\mu_1(\Omega_i) = \mu_2(\Omega_i) < \infty$ for all i. Then $\mu_1(B) = \mu_2(B)$ for all $B \in \sigma(\mathcal{F}_0)$.

The theorem states that if two measures on $\sigma(\mathscr{F}_0)$ agree on a field $\mathscr{F}_0$ that generates $\sigma(\mathscr{F}_0)$ and also agree on a sequence of disjoint sets that cover the entire sample space, then they agree on $\sigma(\mathscr{F}_0)$.

Proof Let S be a set in $\mathscr{F}_0$ such that $\mu_1(S) = \mu_2(S) < \infty$. Consider the collection of subsets

$$\mathscr{M}_S \triangleq \{Q \in \sigma(\mathscr{F}_0) : \mu_1(S \cap Q) = \mu_2(S \cap Q)\}.$$

We claim that

(a) $\mathscr{F}_0 \subseteq \mathscr{M}_S$, and
(b) $\mathscr{M}_S$ is a λ-system.

The proofs of (a) and (b) will be given in the second half of the proof. We first show that the measure extension is unique given the claims in (a) and (b).

Since $\mathscr{F}_0$ is certainly a π-system, by the π-λ theorem, we have

$$\mathscr{M}_S \supseteq \sigma(\mathscr{F}_0). \tag{3.3}$$

We apply (3.3) with S replaced by Ω_i, for $i = 1, 2, 3, \ldots$. This yields $\sigma(\mathscr{F}_0) \subseteq \mathscr{M}_{\Omega_i}$ for all i.

Let E be any event in $\sigma(\mathscr{F}_0)$. We can write E as

$$E = E \cap \Omega = E \cap \left(\biguplus_{i=1}^{\infty} \Omega_i \right) = \biguplus_{i=1}^{\infty} (E \cap \Omega_i).$$

Since μ_1 and μ_2 are measure functions, they enjoy the property of σ-additivity. Hence,

$$\mu_k(E) = \sum_{i=1}^{\infty} \mu_k(E \cap \Omega_i)$$

for $k = 1, 2$.

Since $E \in \sigma(\mathscr{F}_0) \subseteq \mathscr{M}_{\Omega_i}$ for all i, we have

$$\mu_1(E \cap \Omega_i) = \mu_2(E \cap \Omega_i)$$

for all i, by the definition of $\mathscr{M}_{\Omega_i}$. This implies that $\mu_1(E) = \mu_2(E)$ and thus completes the proof of the uniqueness of measure extension.

We now prove (a): Suppose $A \in \mathscr{F}_0$. Then $A \cap S \in \mathscr{F}_0$ because both A and S are in $\mathscr{F}_0$ and $\mathscr{F}_0$ is closed under intersection. By the assumption on μ_1 and μ_2, we have $\mu_1(A \cap S) = \mu_2(A \cap S)$. Therefore, $A \in \mathscr{M}_S$ for all $A \in \mathscr{F}_0$. This proves that $\mathscr{F}_0 \subseteq \mathscr{M}_S$.

To prove (b), we divide the argument into three parts.

(i) We have $\Omega \in \mathcal{M}_S$ because $\mu_1(\Omega \cap S) = \mu_1(S) = \mu_2(S) = \mu_2(\Omega \cap S)$.
(ii) Suppose $A \in \mathcal{M}_S$. Because S can be written as $S = (A \cap S) \uplus (A^c \cap S)$, we have $\mu_k(S) = \mu_k(A \cap S) + \mu_k(A^c \cap S)$ for $k = 1, 2$. Using the assumption that $\mu_1(S) = \mu_2(S) < \infty$, we obtain

$$\mu_1(A^c \cap S) = \mu_1(S) - \mu_1(A \cap S) = \mu_2(S) - \mu_2(A \cap S) = \mu_2(A^c \cap S).$$

Therefore, $A^c \in \mathcal{M}_S$.
(iii) Suppose A_i are mutually disjoint sets in $\mathcal{M}_S$. This means that $\mu_1(A_i \cap S) = \mu_2(A_i \cap S)$ for all i. By σ-additivity, we obtain

$$\mu_1\big(\uplus_i (A_i \cap S)\big) = \mu_2\big(\uplus_i (A_i \cap S)\big) \implies \mu_1\big((\uplus_i A_i) \cap S\big) = \mu_2\big((\uplus_i A_i) \cap S\big).$$

This shows that $\uplus_i A_i$ is in $\mathcal{M}_S$, completing the proof that $\mathcal{M}_S$ is a λ-system. ∎

As an application of the uniqueness result, we prove that a probability measure on $\mathbb{R}$ is uniquely determined by its Stieltjes measure function.

Theorem 3.9
Let P be a probability measure on $(\mathbb{R}, \mathcal{B}(\mathbb{R}))$, and let $F(x)$ be its induced distribution function. If there is another probability Q on $\mathcal{B}(\mathbb{R})$ such that

$$Q((-\infty, x]) = P((-\infty, x]) \qquad \forall x \in \mathbb{R},$$

then $P(B) = Q(B)$ for all $B \in \mathcal{B}(\mathbb{R})$.

Proof We can apply the uniqueness of measure extension theorem to the algebra

$$\mathcal{B}_0 = \{\text{finite disjoint union of sets of the form } (a, b]\}.$$

When $a < b$, the probability of $(a, b]$ with respect to P is given by

$$P((a, b]) = P((-\infty, b]) - P((-\infty, a]) = F(b) - F(a).$$

By extending this to finite disjoint union, we obtain a pre-measure, denoted by μ_0, on the sets in $\mathcal{B}_0$. In fact, if $A = \uplus_{i=1}^n (a_i, b_i]$, with $a_1 < b_1 \le a_2 < b_2 \le \cdots \le a_n < b_n$, then

$$\mu_0(A) = P(A) = \sum_{i=1}^{n}[F(b_i) - F(a_i)].$$

Suppose Q is a probability measure identical to P when restricted to sets in the form $(-\infty, a]$. Then for any $A \in \mathscr{B}_0$, we have

$$Q(A) = \sum_{i=1}^{n}[F(b_i) - F(a_i)] = P(A).$$

Since $\mathscr{B}(\mathbb{R})$ is the σ-field generated by $\mathscr{B}_0$, the extensions Q and P of μ_0 must be the same probability measure by Theorem 3.8. ∎

Problems

3.1. For integer $n = 1, 2, 3, \ldots$, let rem$(n, 3)$ denote the remainder obtained after dividing n by 3. For example, rem$(5, 3) = 2$ and rem$(6, 3) = 0$. For $n \geq 1$, let A_n be the closed interval

$$[\text{rem}(n, 3),\ 5 + e^{-n}]:$$

(a) Find $\limsup_n A_n$ and $\liminf_n A_n$.
(b) Let $F(x)$ be a Stieltjes measure function defined as follows:

$$F(x) = \begin{cases} 0 & \text{if } x < 0 \\ x/4 & \text{if } 0 \leq x < 2 \\ 1 & \text{if } 2 \leq x. \end{cases}$$

We use $F(x)$ to generate a Lebesgue–Stieltjes measure P. Compute the probabilities $P(\limsup_n A_n)$ and $P(\liminf_n A_n)$.

3.2. Let P be a Lebesgue–Stieltjes measure on $\mathbb{R}$ corresponding to a continuous distribution function, i.e., P is a set function on $\mathscr{B}(\mathbb{R})$ and there exists a continuous function $F(x)$ such that $P([a, b]) = F(b) - F(a)$ for all $a < b$, and $P(\mathbb{R}) = 1$:

(a) Show that $P(A) = 0$ for all countable subsets A of $\mathbb{R}$.
(b) If $P(A) = 1$, can we deduce that A is a dense subset of $\mathbb{R}$?

3.3. For $i = 1, 2, 3, \ldots$, let p_i be a nonnegative real number such that $\sum_{i=1}^{\infty} p_i$ is finite. We define a function μ on $2^{\mathbb{N}}$ by

$$\mu(A) \triangleq \begin{cases} \sum_{i \in A} p_i & \text{if } A \text{ has finite cardinality} \\ \infty & \text{if } A \text{ is an infinite set,} \end{cases}$$

for $A \subseteq \mathbb{N}$:

(a) Determine whether μ is finitely additive.
(b) Determine whether μ is σ-additive.

3.4. Suppose $F_1(x)$ and $F_2(x)$ are two Stieltjes measure functions defined on $\mathbb{R}$, and let μ_1 and μ_2 be the corresponding Lebesgue–Stieltjes measures. Show that if μ_1 and μ_2 are the same set functions, then $F_1(x)$ and $F_2(x)$ differ by an additive constant.

3.5. Let $F(x)$ be a Stieltjes measure function that corresponds to a probability measure. Show that the set of points at which $F(x)$ has a jump discontinuity is at most countably infinite.

3.6. Consider a sample space $\Omega = \{a, b, c, d, e, f\}$ and σ-algebra $\mathscr{F} = 2^\Omega$. We define two different probability measures P and Q on this measure space by $P(\omega) = 1/6$ for all $\omega \in \Omega$, and

$$Q(b) = Q(e) = 1/2, \quad Q(a) = Q(c) = Q(d) = Q(f) = 0.$$

Let $\mathscr{C}$ denote the collection of sets

$$\{\{a, b, c\}, \{b, c, d\}, \{c, d, e\}, \{d, e, f\}, \{e, f, a\}, \{f, a, b\}\} :$$

(a) Prove that the σ-algebra generated by $\mathscr{C}$ is the same as $\mathscr{F}$.
(b) Show that $P(A) = Q(A)$ for all $A \in \mathscr{C}$.
(c) Does it violate the uniqueness of measure extension in Theorem 3.8?

3.7. (An alternate definition of λ-system) Show that the following three conditions are equivalent to the definition of λ-system in Definition 3.10:

1. $\Omega \in \mathscr{L}$.
2. If A and B are sets in $\mathscr{L}$ and $A \subseteq B$, then $B \setminus A \in \mathscr{L}$.
3. If $A_1, A_2, A_3, \ldots$ is a sequence of *increasing* sets in $\mathscr{L}$, then $\cup_{i=1}^{\infty} A_i \in \mathscr{L}$.

3.8. Show that a λ-system is a σ-algebra if and only if it is a π-system.

3.9. (Proof of the π-λ theorem) Let $\mathscr{P}$ be a non-empty π-system contained in a λ-system $\mathscr{L}$:

(a) Show that the intersection of all λ-systems included in $\mathscr{L}$ that contain $\mathscr{P}$ is a λ-system. Denote this λ-system by $\mathscr{L}_0$.

(b) Prove that $\mathscr{L}_1 \triangleq \{A \in \mathscr{L}_0 : A \cap B \in \mathscr{L}_0 \text{ for all } B \in \mathscr{P}\}$ is equal to $\mathscr{L}_0$ by showing $\mathscr{L}_1$ is a λ-system.

(c) Prove that $\mathscr{L}_2 \triangleq \{A \in \mathscr{L}_1 : A \cap C \in \mathscr{L}_0 \text{ for all } C \in \mathscr{L}_0\}$ is equal to $\mathscr{L}_0$, by showing $\mathscr{L}_2$ is a λ-system.

(d) Use the previous exercise to show that $\mathscr{L}_0$ is a σ-algebra containing $\mathscr{P}$.

Measurable Functions and Random Variables 4

In measure-theoretic probability, probability mass function and probability density are not the most fundamental concepts. Instead, we introduce the notion of measurable functions to study random variables in a more general setting. By focusing on measurable functions, we can better understand how to combine two or more random variables into a new random variable. Thus, in this chapter, we primarily focus on measurable space, which consists of the sample space and a σ-algebra, without the probability measure.

The collection of random variables is closed under the usual arithmetic operations, meaning that the sum, difference, product, maximum, and minimum of two measurable functions are also measurable. Additionally, we can also show that the point-wise limit of an infinite sequence of measurable functions is measurable. These properties provide a great degree of flexibility for manipulating random variables.

4.1 Measurable Functions

Consider a population of finite size and suppose we want to study the distribution of height and weight in this population. We randomly select a person from the population and measure his/her height and weight. The measurements depend on the sample we have selected and thus vary from sample to sample. We distinguish between the process of randomly sampling a person from the population and the procedure used to measure their height and weight. The former process is random and is described by a probability measure. Once a person is selected, there is no more randomness, and the height and weight are measured using a deterministic procedure.

The body mass index (BMI) is a measure of body fat based on a person's height and weight. It is calculated as the weight in kilograms divided by the height in meters squared and is a deterministic function. In other words, once we know a

© The Author(s), under exclusive license to Springer Nature Switzerland AG 2023
K. Shum, *Measure-Theoretic Probability*, Compact Textbooks in Mathematics,
https://doi.org/10.1007/978-3-031-49830-5_4

person's height and weight, we can compute their BMI without any uncertainty or variability.

However, if we are considering a population of individuals and sampling from that population, then the resulting BMI becomes a random variable. This is because the height and weight of each individual in the population may vary, and thus the resulting BMI values will also vary. But once we have selected a specific individual and measured their height and weight, the resulting BMI value is no longer random or uncertain. It becomes a fixed value that can be computed using the deterministic formula.

This example leads to the following definition of measurable function.

Definition 4.1

Let $(\Omega, \mathscr{F})$ be a measurable space. A function $X : \Omega \to \mathbb{R}$ is called a (real-valued) *measurable function* if for any Borel set B in $\mathscr{B}(\mathbb{R})$, we have

$$X^{-1}(B) \triangleq \{\omega \in \Omega : X(\omega) \in B\} \in \mathscr{F}.$$

We note that the definition of measurable function does not depend on any probability measure, only the σ-algebra is relevant. However, if a probability measure μ is given, and $(\Omega, \mathscr{F}, \mu)$ is a probability space, we refer to X as a *random variable*.

We will use the terms "measurable functions" and "random variables" interchangeably in subsequent discussions. The definition of random variables ensures that we can assign a probability to any set in the form

$$\{\omega \in \Omega : X(\omega) \in B\},$$

where B is any Borel set. If we denote the probability measure by P, then the probability of the above event is written as

$$P(\{\omega \in \Omega : X(\omega) \in B\}).$$

As short-hand notation, we also write it as

$$P(X(\omega) \in B) \text{ or } P(X \in B).$$

In this chapter, we will study the properties of random variables that do not depend on how we assign probability to Borel sets. That is why we use the terminology "measurable function". We will include the assignment of probability later, after we have studied the class of functions to which we can assign probability.

In general, we can define a random variable that takes values in a set Ω' equipped with a σ-field $\mathscr{G}$. The range space Ω' can be Euclidean space $\mathbb{R}^d$, the set of complex numbers $\mathbb{C}$, or the set of infinite binary sequences. When the range is $\mathbb{R}^d$, the measurable function is called a random vector. Similarly, when the range is $\mathbb{C}$, the measurable function is a complex-valued random variable, and when the range is the set of infinite binary sequences, it is called a random sequence.

Definition 4.2

A function f from measurable space $(\Omega, \mathscr{F})$ to measurable space $(\Omega', \mathscr{G})$ is called $(\mathscr{F}, \mathscr{G})$-*measurable* if

$$f^{-1}(B) \triangleq \{\omega \in \Omega : f(\omega) \in B\}$$

is $\mathscr{F}$-measurable for all $B \in \mathscr{G}$. When the σ-field $\mathscr{G}$ in the codomain Ω' is understood from the context, we say that the function f is $\mathscr{F}$-*measurable*. If both $\mathscr{F}$ and $\mathscr{G}$ are understood, we will just say that the function f is *measurable*. When Ω is the real number line and $\mathscr{F}$ is the Borel algebra, we will refer to an $\mathscr{F}$-measurable function as a *Borel measurable* function.

▶ **Notation** Although f is a mapping from Ω to Ω', we sometime refer to f as a map from the measurable space $(\Omega, \mathscr{F})$ to the measurable space $(\Omega', \mathscr{G})$, when we want to emphasize the associated σ-fields. In probability theory, random variables are generally denoted using capital letters, such as X and Y. Technically speaking, random variables are functions and should be written as $X(\omega)$ and $Y(\omega)$. However, for the sake of convenience, we often use the simplified notation X and Y.

Example 4.1.1 (Measurable Functions)
If $\mathscr{F}$ is the trivial σ-field $\{\emptyset, \Omega\}$, and $\mathscr{G}$ is the σ-field in which all singletons are $\mathscr{G}$-measurable, then an $(\mathscr{F}, \mathscr{G})$-measurable function is a constant function.
    If $\mathscr{F}$ is the power set $\mathscr{P}(\Omega)$, then no matter what $\mathscr{G}$ is, all $(\mathscr{F}, \mathscr{G})$ functions from Ω to Ω' are measurable.

Example 4.1.2 (Random Variable as Dictionary in Python)
To illustrate the concept of a random variable when the sample space is a finite set, we can use a Python dictionary to implement a function that maps elements in a sample space to real numbers. Below is an example Python program.

```
from random import choice
Omega = ['a','b','c','d','e']        # sample space
X = {'a':1, 'b':2, 'c':3, 'd':4, 'e':5}   # random variable X
Y = {'a':0, 'b':3, 'c':2, 'd':-1, 'e':5}  # random variable Y
omega= choice(Omega)   # pick a sample uniformly
print(f"X={X[omega]}, Y={Y[omega]}")  # print the values of X and Y
```

In this program, a random sample is generated using the choice function from the random library. We note that the two random variables are defined before any samples are drawn. Once a sample is generated, the values of X and Y are determined based on data stored in the dictionaries.

▶ **The Choice of σ-Fields in the Domain and Codomain** To maximize flexibility, we would like to choose the σ-field $\mathscr{F}$ in the domain to be as large as possible.

Table 4.1 Codomains of measurable functions

Ω'	$\mathscr{G}$	
$\mathbb{R}$	$\mathscr{B}(\mathbb{R})$	Real-valued random variable
$\bar{\mathbb{R}} = \mathbb{R} \cup \{\pm\infty\}$	$\mathscr{B}(\bar{\mathbb{R}})$	Real-valued random variable with infinity
$\mathbb{R}^d$	$\mathscr{B}(\mathbb{R}^d)$	d-dim. random vector
$\mathbb{C}$	$\mathscr{B}(\mathbb{C})$	Complex-valued random variable
$\{0, 1\}^\infty$	$\mathscr{B}(\{0, 1\}^\infty)$	Infinite random bit stream
$\mathbb{R}^\infty$	$\mathscr{B}(\mathbb{R}^\infty)$	Infinite random sequence

If we choose $\mathscr{F}$ to be the power set $\mathscr{P}(\Omega)$, then every function is measurable. On the other hand, we want $\mathscr{G}$ to be as small as possible, but it cannot be too small. It should contain all the basic events of interests. For example, when the codomain of the measurable function is $\mathbb{R}^n$, we usually take the Borel algebra on $\mathbb{R}^n$ as the σ-field in the codomain but do not necessarily consider complete measure space in the codomain.

By selecting different codomains and the associated σ-algebra, we obtain various types of random variables. Table 4.1 lists a few useful combinations.

We note that when the codomain of a measurable function is the extended real numbers $\bar{\mathbb{R}} \triangleq \mathbb{R} \cup \{\pm\infty\}$, the Borel algebra is generated by the ordinary open sets in $\mathbb{R}$ and the special open sets of the form $(a, \infty]$ and $[-\infty, a)$. Some textbooks refer to $\bar{\mathbb{R}}$-valued measurable functions as *generalized random variables*.

Theorem 4.1
Let $(\Omega, \mathscr{F})$ be a measurable space and let A be a subset of Ω. The indicator function

$$\mathbf{1}_A(\omega) \triangleq \begin{cases} 1 & \text{if } \omega \in A \\ 0 & \text{otherwise.} \end{cases}$$

is $(\mathscr{F}, \mathscr{B}(\mathbb{R}))$-measurable if and only if A is $\mathscr{F}$-measurable.

Proof We consider four cases. Suppose B is a Borel set that contains 1 but not 0. Then the pre-image $\mathbf{1}_A^{-1}(B)$ is equal to the set A. If B is a Borel set that contains 0 but not 1, then $\mathbf{1}_A^{-1}(B) = A^c$. When B contains both 0 and 1, then $\mathbf{1}_A^{-1}(B) = \Omega$. Finally, when B does not contain 0 nor 1, then $\mathbf{1}_A^{-1}(B) = \emptyset$. This exhausts all possibilities of Borel sets B containing 0 or 1.

If A is $\mathcal{F}$-measurable, then the four possible pre-images A, A^c, Ω, and $\emptyset$ are all in $\mathcal{F}$, and hence, the indicator function $\mathbf{1}_A$ is $\mathcal{F}$-measurable. Otherwise, if A is not in $\mathcal{F}$, then the indicator function is not $\mathcal{F}$-measurable. ∎

For example, we can consider the set $\mathbb{Q}$ consisting of all rational numbers in the real number line. It is measurable because it is a countable union of singletons. The indicator function $\mathbf{1}_{\mathbb{Q}}$ is Borel measurable. If C is the Cantor set, the indicator function $\mathbf{1}_C$ is also Borel measurable. However, if V is the Vitali set constructed in Sect. 2.5, the indicator function $\mathbf{1}_V$ is not Borel measurable.

4.2 Composition of Measurable Functions

Given a function f mapping from domain Ω to Ω', the inverse image of a set A in Ω' is defined as the set of elements $x \in \Omega$ such that $f(x) \in A$, i.e.,

$$f^{-1}(A) \triangleq \{x \in \Omega : f(x) \in A\}.$$

We remark that the function f need not be an injective function. The mapping above is defined for any function f.

The inverse image function enjoys several nice properties.

Theorem 4.2

Suppose $f : \Omega \to \Omega'$ and A is any set in Ω'. Then we have:

1.

$$f^{-1}(A^c) = (f^{-1}(A))^c, \tag{4.1}$$

where A^c denotes the complement of A in Ω', and $(f^{-1}(A))^c$ denotes the complement of $f^{-1}(A)$ in Ω.

2. If $(A_i)_{i \in I}$ is any collection of sets in Ω', then

$$f^{-1}(\cap_{i \in I} A_i) = \bigcap_{i \in I} f^{-1}(A_i), \tag{4.2}$$

$$f^{-1}(\cup_{i \in I} A_i) = \bigcup_{i \in I} f^{-1}(A_i). \tag{4.3}$$

The index set I can be uncountably infinite.

3. If h is the composition $g \circ f$, then $h^{-1}(B) = f^{-1}(g^{-1}(B))$ for any subset B of the codomain of g.

Proof We will prove part 1 and leave parts 2 and 3 as exercise.

Let A be any set in Ω'. By definition, $f^{-1}(A)$ consists of all elements x in A such that $f(x)$ is an element in A. Therefore, any element x' that is not in $f^{-1}(A)$ does not satisfy $f(x') \in A$. Equivalently, any element x' in the complement of $f^{-1}(A)$ must satisfy $f(x') \in A^c$. This proves that $(f^{-1}(A))^c \subseteq f^{-1}(A^c)$.

Conversely, let x be any element in $f^{-1}(A^c)$. Then $f(x)$ is an element in A^c, which implies that $f(x)$ cannot be an element in A. Therefore, x must be in the complement of $f^{-1}(A)$. This proves that $f^{-1}(A^c) \subseteq (f^{-1}(A))^c$.

Hence, we have shown that $f^{-1}(A^c) = (f^{-1}(A))^c$.   ∎

We give several simple but important properties of measurable functions. The first one is about the composition of two measurable functions.

> **Theorem 4.3   (Composition of Measurable Functions)**
> *Suppose* $f : (\Omega, \mathscr{F}) \to (\Omega', \mathscr{G})$ *is* $(\mathscr{F}, \mathscr{G})$-*measurable, and* $g : (\Omega', \mathscr{G}) \to (\Omega'', \mathscr{H})$ *is* $(\mathscr{G}, \mathscr{H})$-*measurable. Then the composed function* $h = g \circ f$ *is* $(\mathscr{F}, \mathscr{H})$-*measurable.*

Proof Suppose A is a set in $\mathscr{H}$. Because g is $(\mathscr{G}, \mathscr{H})$-measurable, the pre-image $g^{-1}(A)$ is in $\mathscr{G}$. Because f is $(\mathscr{F}, \mathscr{G})$-measurable, we have $f^{-1}(g^{-1}(A))$ in $\mathscr{F}$. The proof is completed by noting that $h^{-1}(A) = f^{-1}(g^{-1}(A))$.   ∎

Some analysis textbooks state that the composition of two measurable functions may not be measurable. However, this statement refers specifically to the Lebesgue σ-field, which is larger than the Borel σ-field used in the definition of measurability in Theorem 4.3. The Lebesgue σ-field on $\mathbb{R}$, which we denote by $\mathscr{L}(\mathbb{R})$, is a complete measure constructed using the outer measure and is much larger than the Borel algebra. The statement in those analysis textbook is in fact "if f and g are $(\mathscr{L}(\mathbb{R}), \mathscr{B}(\mathbb{R}))$-measurable, then the composition $g \circ f$ may not be $(\mathscr{L}(\mathbb{R}), \mathscr{B}(\mathbb{R}))$-measurable". This makes sense because, for any Borel set B, the inverse image $f^{-1}(B)$ in general is a set in $\mathscr{L}(\mathbb{R})$ and may not be in $\mathscr{B}(\mathbb{R})$. However, the σ-fields in Theorem 4.3 are explicitly given, and there is no such mismatch.

As an example, let N denote a null set in $\mathbb{R}$ that is not Borel measurable, such as a non-Borel measurable subset of the Cantor set (See Sect. 3.4). Because the Lebesgue σ-algebra is complete with respect to the Lebesgue measure, the null set N is in $\mathscr{L}(\mathbb{R})$, but not in $\mathscr{B}(\mathbb{R})$. Suppose $f(x) = x^2$ is the square function, and $g(x) = \mathbf{1}_N(x)$ be the indicator function of N. Then there is no guarantee that the composition $g \circ f$ is $(\mathscr{L}(\mathbb{R}), \mathscr{B}(\mathbb{R}))$-measurable.

In fact, the compositionality of the measurable functions makes the collection of all measurable spaces a category.

The Category of Measurable Spaces

In category theory, a category is a collection of objects and morphisms that describe the relationships among the objects. We can view probability spaces as the objects of a category, and the random variables as the morphisms between them. Theorem 4.3 states that the category of measurable spaces and measurable functions is closed under composition. By viewing measurable spaces and measurable functions as objects and morphisms in a category, we can apply the tools of category theory to gain deeper understanding of random variables.

The next property provides an effective way to check measurability.

Theorem 4.4

Let $f : (\Omega, \mathscr{F}) \to (\Omega', \mathscr{G})$ be a function, and let $\mathscr{C}$ be a generating set of $\mathscr{G}$, i.e., $\sigma(\mathscr{C}) = \mathscr{G}$. Then, f is $(\mathscr{F}, \mathscr{G})$-measurable if and only if $f^{-1}(A) \in \mathscr{F}$ for all $A \in \mathscr{C}$.

Proof The "only if" part follows directly from the definition of measurable functions. To prove the other direction, suppose $f^{-1}(A) \in \mathscr{F}$ for all $A \in \mathscr{C}$. Let

$$\mathscr{H} \triangleq \{B \subseteq \Omega' : f^{-1}(B) \in \mathscr{F}\}.$$

We claim that $\mathscr{H}$ is a σ-field containing $\mathscr{C}$.

To see this, note that Ω' is in $\mathscr{H}$ since $f^{-1}(\Omega') = \Omega$. Moreover, if $B \in \mathscr{H}$, then $f^{-1}(B) \in \mathscr{F}$, and by (4.1), we have $f^{-1}(B^c) = (f^{-1}(B))^c$, proving that $f^{-1}(B^c)$ is in $\mathscr{F}$, and hence, B^c is in $\mathscr{H}$. Similarly, by applying (4.2) and (4.3), we can show that $\mathscr{H}$ is closed under countable intersection and countable union. This proves that $\mathscr{H}$ is a σ-field.

By assumption $\mathscr{H}$ contains all sets in $\mathscr{C}$. Hence, $\mathscr{C} \subseteq \sigma(\mathscr{C}) \subseteq \mathscr{H}$. Since $\mathscr{G}$ is the same as $\sigma(\mathscr{C})$, we get $\mathscr{G} = \sigma(\mathscr{C}) \subseteq \mathscr{H}$, i.e., every set in $\mathscr{G}$ is in $\mathscr{H}$. By the definition of $\mathscr{H}$, we obtain $f^{-1}(A) \in \mathscr{F}$ for any set A in $\mathscr{G}$. ∎

We note that this is an indirect proof, as there is no explicit description of all Borel sets. However, this result provides a large supply of measurable functions. For concreteness, we present the following theorem for functions with domain $\mathbb{R}^m$ and codomain $\mathbb{R}^n$. The result holds for more general topological spaces.

Theorem 4.5

A continuous function $f : \mathbb{R}^m \to \mathbb{R}^n$ is $(\mathscr{B}(\mathbb{R}^m), \mathscr{B}(\mathbb{R}^n))$-measurable. In particular, a continuous real-valued function $f : \mathbb{R}^m \to \mathbb{R}$ is measurable.

Proof Recall that $\mathscr{B}(\mathbb{R}^n)$ is the smallest σ-algebra containing all open balls in $\mathbb{R}^n$ (see Definition 2.8). Since f is continuous, the inverse image of any open ball under

f is an open set in $\mathbb{R}^m$, which belongs to $\mathscr{B}(\mathbb{R}^m)$. Hence, by Theorem 4.4, f is $(\mathscr{B}(\mathbb{R}^m), \mathscr{B}(\mathbb{R}^n))$-measurable.   ∎

This theorem says that the class of measurable functions is generally larger than the class of continuous functions.

4.3   Operations with Measurable Functions

Measurable functions can be combined using basic operations, and the resulting function is also measurable. In addition to the usual arithmetic operations, it is also possible to take the limsup and liminf of a sequence of measurable functions. These operations preserve measurability and are useful for constructing new measurable functions from existing ones.

Theorem 4.6

If $f, g : \Omega \to \mathbb{R}$ are $\mathscr{F}$-measurable, then $f + g$, $f - g$, $c \cdot f$, and f/g are measurable, where c is a constant, and in f/g, we assume that $g(\omega)$ is nonzero for all ω.

Proof Suppose f and g are measurable functions. Consider

$$(\Omega, \mathscr{F}) \xrightarrow{\alpha(\omega)} (\mathbb{R}^2, \mathscr{B}(\mathbb{R}^2)) \xrightarrow{\beta(x,y)} (\mathbb{R}, \mathscr{B}(\mathbb{R}))$$

with $\alpha(\omega)$ defined as $(f(\omega), g(\omega))$, and $\beta(x, y)$ defined as $x + y$. The composite function is $f(\omega) + g(\omega)$. To show that the first function α is measurable, we note that $\mathscr{B}(\mathbb{R}^2)$ can be generated by open rectangles in the form $(a, b) \times (c, d)$ (Theorem 2.8), and

$$\alpha^{-1}((a, b) \times (c, d)) = f^{-1}((a, b)) \cap g^{-1}((c, d)).$$

Since both $f^{-1}((a, b))$ and $g^{-1}((c, d))$ are $\mathscr{F}$-measurable (by the assumptions on f and g), $\alpha^{-1}((a, b) \times (c, d))$ is $\mathscr{F}$-measurable.

On the other hand, the addition function $\beta(x, y) = x + y$ is Borel measurable, because, for any constant b, the pre-image, $\{(x, y) : x + y < b\}$, can be expressed as a countable union of open sets

$$\bigcup_{\substack{p, q \in \mathbb{Q} \\ p+q<b}} (\{x < p\} \cap \{y < q\}).$$

As a result, the function $\beta(x, y) = x + y$ is measurable by Theorem 4.4. By Theorem 4.3, the composition $\beta \circ \alpha$ is measurable. This proves that $f + g$ is measurable.

Similarly, we can show that the difference, product, and quotient of measurable functions are measurable. To do this, we first show that the linear function $x \mapsto ax$, the square function $x \mapsto x^2$, and the inverse function $x \mapsto 1/x$ are measurable. For any x in $\mathbb{R}$, we check that

$$\{x : ax < b\} = \begin{cases} \{x : x < b/a\} & \text{if } a > 0, \\ \{x : x > b/a\} & \text{if } a < 0, \end{cases}$$

$$\{x : x^2 < b\} = \begin{cases} \{x : -\sqrt{b} < x < \sqrt{b}\} & \text{if } b > 0, \\ \emptyset & \text{otherwise,} \end{cases}$$

and

$$\{x : 1/x < b\} = \begin{cases} \{x : x < 0\} \cup \{x : 1/b < x\} & \text{if } b > 0, \\ \{x : x < 0\} & \text{if } b = 0, \\ \{x : 1/b < x < 0\} & \text{if } b < 0. \end{cases}$$

Since the sets on the right-hand sides of the displayed equations are all open sets, the sets on the left-hand sides are also open sets. By Theorem 4.4, the linear function, square function, and inverse function are Borel measurable.

We now see that the product function $\beta(x, y) = xy$ is Borel measurable by writing xy as

$$xy = \frac{1}{2}((x + y)^2 - x^2 - y^2).$$

Likewise, the quotient function x/y is Borel measurable, because it can be written as

$$x/y = x(y^{-1}).$$

We can use a similar proof technique as in the proof that $f + g$ is measurable to show that $f \cdot g$, f/g, and $c \cdot f$ are measurable for any measurable functions f and g, and any constant c. ∎

It is convenient to include the special symbols ∞ and $-\infty$ and to work in the extended real number system $\bar{\mathbb{R}} = \mathbb{R} \cup \{\pm\infty\}$. We have the following convention for working with the ∞ symbol.

▶ **Convention Involving Infinity in Probability Theory** The symbols

$$\infty - \infty, \ (-\infty) + \infty, \ \infty/\infty, \ \infty/(-\infty), \ (-\infty)/\infty, \ \text{and} \ (-\infty)/(-\infty)$$

are not defined. However, by convention, we define $0 \cdot \infty$ and $0 \cdot (-\infty)$ to be equal to 0. The other rules for working with infinity follow from common sense. For example,

$$x \cdot \infty = \begin{cases} \infty & \text{if } x > 0 \\ -\infty & \text{if } x < 0 \\ 0 & \text{if } x = 0. \end{cases}$$

These conventions are useful when dealing with limits of functions. The following is an example of measurable function that may take ∞ as its value.

Example 4.3.1 (A Random Variable Whose Value May Be ∞)
Suppose X, Y, and Z are three real-valued random variables defined on a common probability space. Using the machinery we develop so far, such as the continuity of the square root function and the fact that the sum and product of measurable functions are measurable, we can see that $\sqrt{X^2 + Y^2 + Z^2}$ is measurable. The inverse $(X^2 + Y^2 + Z^2)^{-1/2}$ may take ∞ as its value, and this happen when X, Y, and Z are all zero. In this case, we can define $(X^2 + Y^2 + Z^2)^{-1/2}$ as ∞.

Recall that the *supremum* of a set A of real numbers, denoted by sup A, is defined as the smallest upper bound of A. More precisely, sup A is the real number $r \in \mathbb{R}$ that satisfies two conditions: (i) $r \geq a$ for all $a \in A$ and (ii) if $u \geq a$ for all $a \in A$, then $u \geq r$. The first condition says that r is an upper bound of A, and the second condition says that r is the smallest one. If the set A is unbounded from above, we set sup $A = \infty$. Thus, sup A may take a value in the extended real number system. The existence of supremum for any set $A \subseteq \mathbb{R}$ is equivalent to the completeness of the real number system with respect to the absolute value function, and hence, it can be regarded as an axiom of $\mathbb{R}$.

Dually, the *infimum* of a set $A \subseteq \mathbb{R}$, denoted by inf A, is the largest lower bound of A. It is the real number $s \in \mathbb{R}$ that satisfies two conditions: (i) $s \leq a$ for all $a \in A$, and (ii) if $v \leq a$ for all $a \in A$, then $v \leq s$. The first condition says that s is a lower bound of A, and the second condition says that s is the largest one. When A is not bounded from below, we set inf $A = -\infty$.

When $A = \{a_1, a_2, a_3, \ldots\}$ is a countable set, we use the notation

$$\sup_i a_i \triangleq \sup\{a_i : i = 1, 2, 3, \ldots\}, \quad \text{and} \quad \inf_i a_i \triangleq \inf\{a_i : i = 1, 2, 3, \ldots\}.$$

Theorem 4.7
Suppose $f_i : \Omega \to \bar{\mathbb{R}}$ is a measurable function, for $i = 1, 2, 3, \ldots$. Then the functions $X(\omega) \triangleq \sup_i f_i(\omega)$ and $Y(\omega) \triangleq \inf_i f_i(\omega)$ are measurable.

Proof For any fixed real number r, using the definition that the supremum is the smallest upper bound, we obtain

$$\{\omega : \sup_i f_i(\omega) \le r\} = \bigcap_{i=1}^{\infty}\{\omega : f_i(\omega) \le r\}.$$

Because the right-hand side is a countable intersection of measurable sets, the left-hand side is also measurable. By Theorem 4.4, the supremum $\sup_i f_i$ is measurable. We can prove that $\inf_i f_i(\omega)$ is measurable from the relation $\inf_i f_i(\omega) = -\sup_i(-f_i(\omega))$. ∎

Given a sequence of real numbers $(a_i)_{i=1}^{\infty}$, the limsup and liminf of this sequence are defined as

$$\limsup_{i\to\infty} a_i \triangleq \inf_{j\ge 1}\left(\sup_{k\ge j} a_k\right),$$

$$\liminf_{i\to\infty} a_i \triangleq \sup_{j\ge 1}\left(\inf_{k\ge j} a_k\right).$$

In general, the limsup and liminf of a sequence have the following characterizations:

(a) $\limsup_k a_k$ is the smallest real number r such that for any $\epsilon > 0$, there is a sufficiently large integer N such that $a_k < r + \epsilon$ for all $k \ge N$.
(b) $\liminf_k a_k$ is the largest real number s such that for any $\epsilon > 0$, there is a sufficiently large integer N such that $a_k > s + \epsilon$ for all $k \ge N$.
    For any sequence $(a_i)_{i=1}^{\infty}$, it is generally true that

$$\liminf_{i\to\infty} a_i \le \limsup_{i\to\infty} a_i.$$

The inequality becomes an equality when $\lim_{i\to\infty} a_i$ exists.

Example 4.3.2 (limsup and liminf)
Consider the numerical sequence $(a_i)_{i=1}^{\infty}$ defined by

$$a_i \triangleq \begin{cases} 1 + \frac{1}{i} & \text{if } i \text{ is odd,} \\ -1 - \frac{1}{i} & \text{if } i \text{ is even.} \end{cases}$$

This sequence may be regarded as the interleaving of two sequences that converge to 1 and -1, respectively. However, the sequence $(a_i)_{i=1}^{\infty}$ itself does not converge. The limsup and liminf of this sequence are 1 and -1, respectively.

Theorem 4.8
Suppose $f_i : \Omega \to \bar{\mathbb{R}}$ is a measurable function, for $i = 1, 2, 3, \ldots$ .. The two functions

$$X(\omega) \triangleq \limsup_i f_i(\omega) \text{ and } Y(\omega) \triangleq \liminf_i f_i(\omega)$$

are measurable. Furthermore, if $\lim_{i \to \infty} f_i(\omega)$ exists for all $\omega \in \Omega$, then the function

$$Z(\omega) = \lim_{i \to \infty} f_i(\omega)$$

is measurable.

Proof The limsup and liminf of measurable functions are measurable, because

$$\limsup_{i \to \infty} f_i(\omega) = \inf_{j \geq 1} \sup_{k \geq j} f_k(\omega),$$

$$\liminf_{i \to \infty} f_i(\omega) = \sup_{j \geq 1} \inf_{k \geq j} f_k(\omega),$$

which can be computed in terms of supremum and infimum. From the previous theorem, we know that the supremum and infimum of measurable functions are measurable.

If the point-wise limit of f_i exists, then the limsup and liminf of f_i coincide. Thus,

$$\lim_{i \to \infty} f_i(\omega) = \limsup_{i \to \infty} f_i(\omega) = \liminf_{i \to \infty} f_i(\omega)$$

is measurable.   ■

The operations discussed in this section are sufficient for most applications. For instance, if X and Y are random variables, we can apply the theorems proved in this section to see that

$$Z = \max(X, Y) + 3 \sin(X) \exp(Y)$$

is also a random variable. In general, the cdf $F_Z(z) = P(Z \leq z)$ may be difficult to compute and may not have a closed-form expression. However, we can at least guarantee that $P(Z \leq z)$ is well-defined because we know that $\{\omega : Z(\omega) \leq z\}$ is a measurable set.

4.4   Complex-Valued Random Variables

A *complex number* z is an expression in the form $a + bi$, where a and b are real numbers and i is a symbol that satisfies $i^2 = -1$. The *real part a* of z is denoted by $\text{Re}(z)$ and the *imaginary part b* by $\text{Im}(z)$. The set of all complex numbers is denoted by $\mathbb{C}$. We can identify the complex numbers with the two-dimensional space $\mathbb{R}^2$, where the real part corresponds to the x-axis and the imaginary part corresponds to the y-axis. Using this identification, the Borel algebra of $\mathbb{C}$ is the same as the Borel algebra of $\mathbb{R}^2$. For example, we can generate the Borel algebra of $\mathbb{C}$ by open rectangles in the form

$$\{x + yi \in \mathbb{C} : a < x < b, \, c < y < d\},$$

where $a < b$ and $c < d$ are real numbers.

The addition of complex numbers is equivalent to vector addition.

$$(a + bi) + (c + di) = (a + c) + (b + d)i.$$

The absolute value (also known as *modulus* or *magnitude*) of z is the length of the vector (a, b) and is defined as $|z| = \sqrt{a^2 + b^2}$. The absolute value satisfies the triangle inequality,

$$|z_1 + z_2| \le |z_1| + |z_2|.$$

We perform complex multiplication by expanding

$$(a + bi)(c + di) = ac + ibc + iad + i^2bd = (ac - bd) + i(bc + ad).$$

The *conjugate* of z denoted by $(a + bi)^* \triangleq a - bi$ is the reflection of z along the real axis.

Suppose we have a probability space $(\Omega, \mathcal{F}, P)$. A *complex random variable* is a complex-valued function from Ω to $\mathbb{C}$ such that the inverse image $X^{-1}(B)$ is in $\mathcal{F}$ for all B in $\mathcal{B}(\mathbb{C})$. A complex random variable $Z(\omega)$ can be represented as $X(\omega) + iY(\omega)$, where $X(\omega)$ and $Y(\omega)$ are real-valued functions. We can see that $X(\omega)$ and $Y(\omega)$ are measurable by observing that the real and imaginary parts of a complex number z can be obtained as $\text{Re}(z) = (z + z^*)/2$ and $\text{Im}(z) = (z - z^*)/(2i)$, respectively. Furthermore, since $Z(\omega)^*$ is the composition of the measurable function Z and a continuous reflection, we have

$$Z(\omega) \text{ is measurable} \implies Z(\omega)^* \text{ is measurable.}$$

Consequently, the real part $\text{Re}(Z) = (Z + Z^*)/2$ and imaginary part $\text{Im}(Z) = (Z - Z^*)/(2i)$ are both measurable if Z is measurable and *vice versa*.

We can represent a complex number $z = a + bi$ in polar form as

$$z = r\cos\theta + ir\sin\theta,$$

where $r = \sqrt{a^2 + b^2}$ is called the *radius*, and θ is the *argument* or *phase*. We denote the radius and argument of a complex number z by $|z|$ and $\arg(z)$, respectively. Note that the argument $\arg(z)$ has a 2π ambiguity when $r > 0$. When $r = 0$, the argument is undefined. When two complex numbers are given in polar form, we can compute their product as follows:

$$(r_1\cos\theta_1 + ir_1\sin\theta_1)(r_2\cos\theta_2 + ir_2\sin\theta_2) = r_1 r_2(\cos(\theta_1 + \theta_2) + i\sin(\theta_1 + \theta_2)).$$

Using Euler's formula

$$e^{i\theta} = \cos\theta + i\sin\theta,$$

complex multiplication can be simplified as

$$(r_1 e^{i\theta_1})(r_2 e^{i\theta_2}) = r_1 r_2 e^{i(\theta_1 + \theta_2)}.$$

We can represent a complex-valued random variable $Z(\omega)$ in polar form as

$$Z(\omega) = R(\omega)e^{i\Theta(\omega)},$$

where $R(\omega)$ is a nonnegative function, and $\Theta(\omega)$ takes values in the quotient group $\mathbb{R}/(2\pi\mathbb{Z})$. Geometrically, we can visualize this quotient group as a circle with circumference 2π.

Example 4.4.1 (Unit Disc in Complex Plane)
Suppose $Z(\omega)$ is a complex random variable that is uniformly distributed inside the unit disc

$$D \triangleq \{z \in \mathbb{C} : |z| < 1\}$$

in the complex plane. By symmetry, the argument $\Theta(\omega)$ is uniformly distributed between 0 and 2π. The radius is distributed between 0 and 1 with cumulative distribution function

$$F(r) \triangleq \Pr(|Z(\omega)| < r) = \frac{2\pi r^2}{2\pi(1)^2} = r^2,$$

for $0 \leq r \leq 1$.

Example 4.4.2 (Eigenvalue of Random Matrix)
Consider 2×2 skew-symmetric matrix

$$\begin{bmatrix} 0 & X(\omega) \\ -X(\omega) & 0 \end{bmatrix},$$

where $X(\omega)$ is a real-valued random variable distributed according to the standard Gaussian distribution. For any fixed $\omega \in \Omega$, we can compute the eigenvalues of this random matrix, and they are the roots of the characteristic polynomial

$$\det \begin{bmatrix} -\lambda & X(\omega) \\ -X(\omega) & -\lambda \end{bmatrix} = \lambda^2 + X(\omega)^2.$$

Hence, the eigenvalues are $iX(\omega)$ and $-iX(\omega)$. The two eigenvalues are complex random variables.

Example 4.4.3 (Complex Gaussian Random Variable)

The standard complex Gaussian random variable is defined as a complex random variable whose real part and imaginary part are independent real-valued normal random variables with mean 0 and variance 1/2. We use the notation $Z \sim CN(0, 1)$ if Z is a complex-valued random variable distributed according to the standard complex Gaussian distribution. The variance of Z is equal to $E[\text{Re}(Z)^2 + \text{Im}(Z)^2] = 1$.

A complex Gaussian random variable Z is *circularly symmetric*, which means that its probability distribution is invariant under rotations in the complex plane, i.e., for any real number θ, the complex random variable $e^{i\theta}Z$ has the same distribution as Z. The complex Gaussian distribution is commonly used to model additive noise in wireless communication.

Using Python, we can generate $Z \in CN(0, 1)$ using the following commands:

```
from numpy import random
Z = random.normal(0,1/2) + random.normal(0,1/2)*1j
print(Z)   # print the random sample
```

A sample output is

```
(-0.2916888782004312+0.4270105512204956j)
```

Problems

4.1. Let $(\Omega, \mathscr{F})$ be a measurable space and f be a function from Ω to $\mathbb{R}$. Show that if one of the following conditions hold:

- $f^{-1}((\infty, a)) \in \mathscr{F}$ for all $a \in \mathbb{R}$
- $f^{-1}((\infty, a]) \in \mathscr{F}$ for all $a \in \mathbb{R}$

then f is $(\mathscr{F}, \mathscr{B}(\mathbb{R}))$-measurable.

4.2. Suppose $f, g : \Omega \to \mathbb{R}$ are real-valued random variables. Prove that

$$|f|, \; \max(f, g), \; \min(f, g)$$

are random variables.

4.3. Find an example of function f defined on a measure space such that f is not measurable, but $|f|$ is measurable.

4.4. Let $Z(\omega)$ be a complex random variable and α be a complex constant. Show that $\alpha Z(\omega)$ is a complex random variable.

4.5. Let $(\Omega, \mathscr{F})$ be a measurable space and $f_1, f_2, \ldots, f_n$ are measurable functions from $(\Omega, \mathscr{F})$ to $(\mathbb{R}, \mathscr{B}(\mathbb{R}))$. Prove that for any continuous function $g : \mathbb{R}^n \to \mathbb{R}$, the function $g(f_1(\omega), f_2(\omega), \ldots, f_n(\omega))$ is measurable with respect to the σ-algebra $\mathscr{F}$.

4.6. Define a collection of subsets in $\mathbb{R}$ by $\mathscr{G} \triangleq \{\cup_{n \in S}(n, n+1] : S \in \mathbb{Z}\}$:

(a) Show that $\mathscr{G}$ is a σ-algebra.
(b) Show that a $\mathscr{G}$-measurable function f is constant on the interval $(n, n+1]$ for each integer n.

4.7. (Borel sets in $\mathbb{R}^2$) Take the two-dimensional Euclidean space $\mathbb{R}^2$ as the sample space, and consider the σ-algebra $\mathscr{F}$ generated by open rectangles

$$(a, b) \times (c, d) = \{(x, y) \in \mathbb{R} : a < x < b, \ c < y < d\},$$

where a, b, c, and d are rational numbers.
   Show that the following types of sets are $\mathscr{F}$-measurable:

(a) Cartesian product $(p, q) \times (r, s)$, with $p, q, r, s \in \mathbb{R}$,

(b) Cartesian product $(-\infty, a) \times (-\infty, b)$, with $a, b \in \mathbb{Q}$,

(c) Open half plane $\{(x, y) \in \mathbb{R}^2 : ax + by < c\}$, with $a, b, c \in \mathbb{R}$.

4.8.

(a) Find $\limsup_{n \to \infty} \sin(n\pi/3)$ and $\liminf_{n \to \infty} \sin(n\pi/3)$.
(b) For $k \geq 1$, define

$$a_k = \begin{cases} \log(k) & \text{if } k \text{ is odd} \\ 1/k & \text{if } k \text{ is even.} \end{cases}$$

Find the limsup and liminf of the sequence $(a_k)_{k=1}^{\infty}$.

4.9. Suppose a_k and b_k are real numbers satisfying $a_k \leq b_k$ for $k = 1, 2, 3, \ldots$. Prove:

(a) $\liminf_k a_k \leq \liminf_k b_k$.
(b) $\limsup_k a_k \leq \limsup_k b_k$.

4.10. Show that for any numerical sequence $(a_k)_{k=1}^{\infty}$,

$$\liminf_k a_k \leq \limsup_k a_k.$$

4.11. Suppose $(a_k)_{k=1}^{\infty}$ is a sequence of nonnegative real numbers. Prove that if $\limsup_k a_k = 0$, then $\lim_{k \to \infty} a_k = 0$.

4.12. A real-valued function $f(x)$ is said to be *upper semi-continuous* at a point x_0 if for any $y > f(x_0)$, there exists an open neighborhood U of x_0 such that $f(x) < y$ for all x in U. The function $f(x)$ is defined as upper semi-continuous if it is upper semi-continuous at every point in its domain.

Prove that an upper semi-continuous function is Borel measurable, i.e., prove that it is $(\mathscr{B}(\mathbb{R}), \mathscr{B}(\mathbb{R}))$-measurable.

4.13. Let A be the set $\{1, 2, 3, 4\}$ and B be $\{a, b, c\}$:

(a) How many distinct functions can be defined from A to B?
(b) Suppose we impose an algebra $\mathscr{F} = \{\emptyset, A, \{1, 2\}, \{3, 4\}\}$ on A, and an algebra $\mathscr{G} = \{\emptyset, B, \{a, b\}, \{c\}\}$ on B. Count the number of $(\mathscr{F}, \mathscr{G})$-measurable functions from A to B.

Statistical Independence

5

Statistical independence is the central concept in probability theory that distinguishes probability theory from real analysis. The notion of independent events is closely related to conditional probability. Consider an event A with probability $P(A)$. If we know another event B has already occurred, we can update our knowledge about event A by revising the probability of A to the conditional probability $\frac{P(A \cap B)}{P(B)}$. If events A and B are independent, then the original probability $P(A)$ should equal the conditional probability $\frac{P(A \cap B)}{P(B)}$, meaning that the likelihood of A does not change given the information from event B.

This chapter presents a definition of independence of random variables that works for all types of random variables and shows that it is consistent with the that is based on probability density function and probability mass function. In the second part of the chapter we discuss the Borel–Cantelli lemmas and their implications.

Throughout this chapter, random variables and events are defined on a probability space $(\Omega, \mathscr{F}, P)$.

5.1   Independence of Two Random Variables

We will start by defining independence for two events.

Definition 5.1

Two events A and B are said to be *independent* if their joint probability equals the product of their individual probabilities, i.e.,

$$P(A \cap B) = P(A)P(B).$$

When $P(B)$ is positive, the above definition of independence is equivalent to saying that the conditional probability of A given B is the same as the probability

© The Author(s), under exclusive license to Springer Nature Switzerland AG 2023
K. Shum, *Measure-Theoretic Probability*, Compact Textbooks in Mathematics,
https://doi.org/10.1007/978-3-031-49830-5_5

of A. However, when $P(B) = 0$, both sides of the equations in Definition 5.1 are zero, trivially satisfying the definition of independence.

We can now define independence for random variables in terms of independence of events.

Definition 5.2

Two random variables X and Y defined on the same probability space are said to be *independent* if the inverse images of any two Borel sets B_1 and B_2 under X and Y, respectively, are independent events, i.e., $X^{-1}(B_1)$ and $Y^{-1}(B_2)$ are independent for all Borel sets B_1 and B_2.

If two random variables X and Y are independent, this implies that the following equation holds for all Borel sets B_1 and B_2:

$$P(X \in B_1, \ Y \in B_2) = P(X \in B_1)P(Y \in B_2).$$

Alternatively, we can formulate the independence of two random variables in terms of the σ-algebras they generate.

Definition 5.3

Given a measurable function X from $(\Omega, \mathscr{F})$ to $(\mathbb{R}, \mathscr{B}(\mathbb{R}))$, the σ-*algebra generated by* X is defined as

$$\sigma(X) \triangleq \{X^{-1}(B) : B \in \mathscr{B}(\mathbb{R})\}.$$

It is easy to check that the collection $\sigma(X)$ of subsets is indeed a sub-σ-algebra of $\mathscr{F}$. Note that the σ-algebra generated by X contains all the information about X that is measurable in the sense of Borel sets.

Definition 5.4

Two sub-σ-algebras $\mathscr{F}_1$ and $\mathscr{F}_2$ of a common σ-algebra $\mathscr{F}$ are said to be *independent* if any set A_1 from $\mathscr{F}_1$ and any set A_2 from $\mathscr{F}_2$ are independent, i.e.,

$$P(A_1 \cap A_2) = P(A_1)P(A_2),$$

for all $A_1 \in \mathscr{F}_1$ and $A_2 \in \mathscr{F}_2$.

We can now re-formulate the notion of independence for random variables in terms of the σ-algebras they generate.

> **Theorem 5.1**
> *Two random variables X and Y are independent if and only if the σ-algebras $\sigma(X)$ and $\sigma(Y)$ generated by X and Y, respectively, are independent.*

If X and Y are independent random variables, it is reasonable to expect that a function of X and a function of Y should also be independent, since they have different sources of randomness. This intuition can be formally justified using the definition of independent random variables.

> **Theorem 5.2**
> *Suppose X and Y are independent random variables defined on a probability space $(\Omega, \mathscr{F}, P)$. If f and g are Borel measurable functions from $\mathbb{R}$ to $\mathbb{R}$, then $f(X)$ and $g(Y)$ are independent random variables.*

Proof We first note that $f(X(\omega))$ and $g(Y(\omega))$ are both $(\mathscr{F}, \mathscr{B}(\mathbb{R}))$-measurable.

To show that $f(X)$ and $g(Y)$ are independent, we need to show that for any two Borel sets B_1 and B_2 in $\mathbb{R}$, the events

$$\{\omega : f(X(\omega)) \in B_1\} \text{ and } \{\omega : g(Y(\omega)) \in B_2\}$$

are independent.

To see this, note that $f^{-1}(B_1)$ and $g^{-1}(B_2)$ are Borel sets in $\mathbb{R}$, because f and g are both Borel measurable. Then, by the independence of X and Y, the two events

$$\{\omega : X(\omega) \in f^{-1}(B_1)\} \text{ and } \{\omega : Y(\omega) \in g^{-1}(B_2)\}$$

are independent. This shows that $X^{-1}(f^{-1}(B_1))$ and $Y^{-1}(g^{-1}(B_2))$ are independent events, and hence $f(X)$ and $g(Y)$ are independent random variables.   ∎

Theorem 5.2 is a powerful result that applies to a wide range of functions f and g, provided that they are Borel measurable. For example, f and g can be any continuous functions or piece-wise continuous functions. This is because continuous functions and piece-wise continuous functions are Borel measurable.

Moreover, the result in Theorem 5.2 can be generalized to the case where X and Y are independent random vectors. Specifically, if $\mathbf{X}$ and $\mathbf{Y}$ are independent random vectors in $\mathbb{R}^m$ and $\mathbb{R}^n$, respectively, and $\mathbf{f} : \mathbb{R}^m \to \mathbb{R}^p$ and $\mathbf{g} : \mathbb{R}^n \to \mathbb{R}^q$ are Borel measurable functions, then $\mathbf{f}(\mathbf{X})$ and $\mathbf{g}(\mathbf{Y})$ are independent random vectors in $\mathbb{R}^p$ and $\mathbb{R}^q$, respectively. The proof is similar to the proof of Theorem 5.2.

The theory we develop in this section works for random variables of any types, including discrete, continuous, singular, or mixed-type random variables. This framework is sufficiently general to address the independence of a discrete random variable and a continuous random variable.

Example 5.1.1 (Independent Random Variables of Two Different Types)
We randomly select a point uniformly from the unit square $[0, 1] \times [0, 1]$ and let the random point be (X, Y), where X and Y are independent uniform random variables. We define random variable U as $\lfloor 10X \rfloor$ and random variable V as $-\ln(Y)$. Random variable U is discrete, while random variable V is continuous. Although the joint distribution function of U and V is undefined, we can still conclude that U and V are independent random variables.

5.2   Independent Random Variables of Discrete Type or Continuous Type

If X and Y are both discrete random variables, their statistical independence can be expressed in terms of their probability mass functions (pmf's). For simplicity of notation, suppose both random variables X and Y take positive integers as their values. In general, the following theorem holds for any countable codomain.

Theorem 5.3
Suppose X and Y are discrete random variables with pmf's

$$p_X(m) \triangleq P(X = m) \text{ and } p_Y(n) \triangleq P(Y = n),$$

respectively, and joint pmf

$$p_{XY}(m, n) \triangleq P(X = m, Y = n)$$

for $m, n \in \mathbb{N}$. Then X and Y are independent if and only if

$$p_{XY}(m, n) = p_X(m)p_Y(n) \tag{5.1}$$

for all $m, n \in \mathbb{N}$.

Proof
($\Rightarrow$) Let m and n be any positive integers. We apply the definition of independence of random variables with $B_1 = \{m\}$ and $B_2 = \{n\}$. We obtain (5.1) directly from the definition of independence of random variables.
($\Leftarrow$) Conversely, consider two subsets B_1 and B_2 of $\mathbb{N}$. We have

$$P(X \in B_1, Y \in B_2) = \sum_{m \in B_1} \sum_{n \in B_2} p_{XY}(m, n).$$

If (5.1) holds, we can compute

$$P(X \in B_1, Y \in B_2) = \sum_{m \in B_1} \sum_{n \in B_2} p_X(m) p_Y(n)$$

$$= \sum_{m \in B_1} p_X(m) \sum_{n \in B_2} p_Y(n)$$

$$= P(X \in B_1) P(Y \in B_2).$$

This proves that X and Y are independent.   ∎

For real-valued random variables, we have the following analogous theorem.

Theorem 5.4

Suppose X and Y are random variables on probability space $(\Omega, \mathscr{F}, P)$. Let

$$F_{XY}(x, y) \triangleq P(X \le x, Y \le y)$$

$$F_X(x) \triangleq P(X \le x)$$

$$F_Y(y) \triangleq P(Y \le y).$$

Then X and Y are independent if and only if

$$F_{XY}(x, y) = F_X(x) F_Y(y) \quad \text{for all } x, y \in \mathbb{R}.$$

Like the previous theorem, the forward part is straightforward. The reverse direction requires the π-λ theorem. We establish a convenient result in the next theorem and apply it to prove Theorem 5.4.

Theorem 5.5

Let $(\Omega, \mathscr{F}, P)$ be a probability space and X and Y be real-valued random variables defined on $(\Omega, \mathscr{F}, P)$. Suppose the Borel algebra $\mathscr{B}(\mathbb{R})$ is generated by a π-system $\mathscr{C}$. Then X and Y are independent if for all $A, B \in \mathscr{C}$,

$$P(X \in A, Y \in B) = P(X \in A) P(Y \in B).$$

Proof Assume $P(X \in A, Y \in B) = P(X \in A) P(Y \in B)$ holds for all $A, B \in \mathscr{C}$.

For any $C \in \mathscr{C}$, define

$$\mathscr{L}_C \triangleq \{A \in \mathscr{B}(\mathbb{R}) : P(X \in A, Y \in C) = P(X \in A)P(Y \in C)\}.$$

We have $\mathscr{C} \subseteq \mathscr{L}_C$ by assumption, and we want to show that $\mathscr{L}_C$ is a λ-system. (See Def. 3.10 for the definition of λ-system.)

(i) Since $P(X \in \Omega, Y \in C) = P(Y \in C) = P(X \in \Omega)P(Y \in C)$, we have $\Omega \in \mathscr{L}_C$.

(ii) Suppose A is in $\mathscr{L}_C$. Then,

$$P(X \in A^c, Y \in C)$$
$$= P(Y \in C) - P(X \in A, Y \in C)$$
$$= P(Y \in C) - P(X \in A)P(Y \in C) \qquad \text{(because } A \in \mathscr{L}_C)$$
$$= P(X \in A^c)P(Y \in C).$$

Therefore $A^c \in \mathscr{L}_C$.

(iii) Suppose A_i is a sequence of mutually disjoint sets in $\mathscr{L}_C$ for $i = 1, 2, 3 \ldots$. Then,

$$P(X \in \uplus_i A_i, Y \in C) = \sum_{i=1}^{\infty} P(X \in A_i, Y \in C).$$

Because $A_i \in \mathscr{L}_C$, we can further simplify it to

$$P(X \in \uplus_i A_i, Y \in C) = \sum_{i=1}^{\infty} P(X \in A_i)P(Y \in C)$$
$$= P(Y \in C)P(X \in \uplus_i A_i).$$

Therefore $\uplus_i A_i \in \mathscr{L}_C$. This proves that $\mathscr{L}_C$ is a λ-system for each $C \in \mathscr{C}$.

Since $\mathscr{C} \subseteq \mathscr{L}_C$ by assumption and $\mathscr{C}$ generates $\mathscr{B}(\mathbb{R})$, by applying the π-λ theorem, we obtain

$$\mathscr{B}(\mathbb{R}) = \sigma(\mathscr{C}) \subseteq \mathscr{L}_C.$$

As a result, we have

$$P(X \in B, Y \in C) = P(X \in B)P(Y \in C) \qquad (5.2)$$

for any fixed $C \in \mathscr{C}$ and all Borel sets $B \in \mathscr{B}(\mathbb{R})$.

Now we fix a Borel set $B \in \mathscr{B}(\mathbb{R})$ and define

$$\mathscr{L}_B \triangleq \{C \in \mathscr{B}(\mathbb{R}) : P(X \in B, \, Y \in C) = P(X \in B)P(Y \in C)\}.$$

From what we have proved in (5.2), we have $\mathscr{C} \subseteq \mathscr{L}_B$. By repeating the same steps as in the first part of the proof, we can show that $\mathscr{L}_B$ is a λ-system for any Borel set B. By the π-λ theorem, we have $\mathscr{B}(\mathbb{R}) \subseteq \mathscr{L}_B$ for all $B \in \mathscr{B}(\mathbb{R})$. Hence

$$P(X \in B, Y \in C) = P(X \in B)P(Y \in C)$$

for all Borel sets B and C.   ∎

We prove Theorem 5.4 by applying the previous theorem with

$$\mathscr{C} = \{(-\infty, a] : a \in \mathbb{R}\},$$

which is a π-system that generates the Borel algebra.

Theorems 5.3 and 5.4 show that a more general definition of independence is backward compatible. The definitions of independence commonly taught in a first course of probability are theorems that can be derived from the first principles.

Example 5.2.1 (Independent Gaussian Random Variables)
It is well known that two zero-mean Gaussian random variables, X_1 and X_2, with zero correlation are independent. Without loss of generality, let us assume that X_1 and X_2 have zero mean. By "zero correlation," we mean that $E[X_1 X_2] = 0$. If we set $\rho = 0$ in the joint Gaussian probability density function in (1.3), we obtain

$$f_{\mathbf{X}}(x_1, x_2) = \frac{1}{2\pi\sigma_1\sigma_2} \exp\left(-\frac{x_1^2}{2\sigma_1^2} - \frac{x_2^2}{2\sigma_2^2}\right),$$

which can be factorized as the product of two Gaussian probability density functions

$$f_{\mathbf{X}}(x_1, x_2) = \frac{1}{\sqrt{2\pi}\sigma_1} e^{-x_1^2/(2\sigma_1^2)} \frac{1}{\sqrt{2\pi}\sigma_2} e^{-x_2^2/(2\sigma_2^2)}.$$

By Theorem 5.4, we conclude that X_1 and X_2 are independent, which means that for any Borel sets B_1 and B_2 in $\mathbb{R}$, we have

$$P(X_1 \in B_1, \, X_2 \in B_2) = P(X_1 \in B_1) \cdot P(X_2 \in B_2).$$

While uncorrelatedness is necessary for independence, it is a much weaker condition. When the joint probability distribution is not Gaussian, two random variables can be uncorrelated while still being independent.

It is important to note that the term "jointly" is also key here. It is possible to construct two dependent and uncorrelated random variables that are Gaussian but not jointly Gaussian. For example, consider two independent random variables: $X \in N(0, 1)$, a standard Gaussian random variable, and U, a discrete random variable that takes the values 1 and -1 with equal probability. We define a new random variable Y as the product of U and X, i.e., $Y = UX$. It follows that Y also follows a standard Gaussian distribution. However, the correlation between X and Y is zero. This illustrates that if two Gaussian random variables have zero mean and zero correlation but are not jointly Gaussian, then they could be dependent.

Example 5.2.2 (Function of Dependent Random Variables Can Be Independent)
The converse of Theorem 5.2 is false. We can construct dependent random variables whose squares
are independent. Consider the example of jointly distributed random variables X and Y, defined by
the joint probability density function

$$f_{XY}(x, y) = \frac{1}{4}(1 + xy)$$

for $(x, y) \in [-1, 1] \times [-1, 1]$. One can check that the marginal pdf's $f_X(x)$ and $f_Y(y)$ are
both constant $1/2$ over the range of X and the range of Y, respectively. Hence, X and Y are not
independent. However, we can show that

$$P(X^2 \le x, \ Y^2 \le y) = \sqrt{x}\sqrt{y} = P(X^2 \le x) \cdot P(Y^2 \le y)$$

for $-1 \le x, y \le 1$. By Theorem 5.4, the random variables X^2 and Y^2 are independent.

5.3   Independence of More Than Two Random Variables

We generalize the notion of independence to infinitely many random variables.

Definition 5.5

For any positive integer n, events $A_1, A_2, \ldots, A_n$ in Ω are said to be *mutually
independent*, or simply *independent*, if for any $2 \le k \le n$ and any k distinct
indices $i_1, i_2, \ldots, i_k \in \{1, 2, \ldots, n\}$,

$$P\left(\bigcap_{j=1}^{k} A_{i_j}\right) = \prod_{j=1}^{k} P(A_{i_j}). \tag{5.3}$$

For example, when we say that three events A_1, A_2, and A_3 are independent, we
mean that

$$P(A_1 \cap A_2) = P(A_1)P(A_2),$$
$$P(A_1 \cap A_3) = P(A_1)P(A_3),$$
$$P(A_2 \cap A_3) = P(A_2)P(A_3), \quad \text{and}$$
$$P(A_1 \cap A_2 \cap A_3) = P(A_1)P(A_2)P(A_3).$$

The notion of pairwise independence is a much weaker condition compared with
mutually independence.

Definition 5.6

The n events $A_1, A_2, \ldots, A_n$ in Ω are said to be *pairwise independent* if (5.3)
holds for $k = 2$.

Mutually independence certainly implies pairwise independence, but in general the reverse implication does not hold.

Example 5.3.1 (Pairwise Independent But Not Mutually Independent Events)

Suppose $\Omega = \{a, b, c, d\}$ and each of the outcomes a, b, c, and d has probability 1/4. Let E_1 be the event $\{a, d\}$, E_2 be $\{b, d\}$, and E_3 be $\{c, d\}$. Each of the events E_1, E_2, and E_3 has probability 1/2 and contains the outcome d. The intersection of any pair of them is $\{d\}$ and hence has probability 1/4. The events E_1, E_2, and E_3 are thus pairwise independent. However, they are not mutually independent, because $P(E_1 \cap E_2 \cap E_3) = 1/4$, which is not equal to the product $P(E_1)P(E_2)P(E_3) = 1/8$.

The following is an elementary property of independent events.

Theorem 5.6
Suppose $B_1, B_2, \ldots, B_n$ are n mutually independent events. For any index i between 1 and n, the events $B_1, B_2, \ldots, B_i^c, \ldots, B_n$ are also mutually independent.

Proof By re-ordering the events we may assume without loss of generality that $i = 1$. We need to show that if we take the complement of B_1, the events $B_1^c, B_2, \ldots, B_n$ are mutually independent.

Take any k indices $j_1, j_2, \ldots, j_k$ between 2 and n, where k can be any integer between 1 and $n - 1$. By the assumption of mutual independence,

$$P(B_{j_1} \cap B_{j_2} \cap \cdots \cap B_{j_k}) = P(B_{j_1})P(B_{j_2}) \cdots P(B_{j_k}),$$
$$P(B_1 \cap B_{j_1} \cap B_{j_2} \cap \cdots \cap B_{j_k}) = P(B_1)P(B_{j_1})P(B_{j_2}) \cdots P(B_{J_k}).$$

On the other hand, we note that

$$(B_1 \cap B_{j_1} \cap B_{j_2} \cap \cdots \cap B_{j_k}) \uplus (B_1^c \cap B_{j_1} \cap B_{j_2} \cap \cdots \cap B_{j_k}) = B_{j_1} \cap B_{j_2} \cap \cdots \cap B_{j_k}.$$

Therefore

$$P(B_1^c \cap B_{j_1} \cap B_{j_2} \cap \cdots \cap B_{j_k})$$
$$= P(B_{j_1} \cap B_{j_2} \cap \cdots \cap B_{j_k}) - P(B_1 \cap B_{j_1} \cap B_{j_2} \cap \cdots \cap B_{j_k})$$
$$= (1 - P(B_1))P(B_{j_1})P(B_{j_2}) \cdots P(B_{j_k}).$$

∎

By applying the previous result multiple times inductively, we have the following extension. Suppose $A_1, A_2, \ldots, A_n$ are n mutually independent events. We take any k distinct indices $i_1, i_2, \ldots, i_k$ in $\{1, 2, \ldots, n\}$, and for each of the chosen indices, let B_{i_j} to be either A_{i_j} or $A_{i_j}^c$. No matter which indices and events are selected, we always have

$$P(B_{i_1} \cap B_{i_2} \cap \cdots \cap B_{i_k}) = P(B_{i_1})P(B_{i_2}) \cdots P(B_{i_k}). \qquad (5.4)$$

We can take a step further and define the meaning of independence of σ-algebras.

Definition 5.7

Suppose $(\Omega, \mathscr{F}, P)$ is a probability space and $\mathscr{F}_1, \mathscr{F}_2, \ldots, \mathscr{F}_n$ are sub-σ-algebras of $\mathscr{F}$. We say that $\mathscr{F}_1, \mathscr{F}_2, \ldots, \mathscr{F}_n$ are *independent* if for any $A_i \in \mathscr{F}_i$, $i = 1, 2, \ldots, n$, we have

$$P(A_1 \cap A_2 \cap \cdots \cap A_n) = P(A_1)P(A_2) \cdots P(A_n).$$

Note that since Ω is in $\mathscr{F}_i$ for all i, the set A_i may equal Ω. Selecting $A_i = \Omega$ in Definition 5.7 is the same as ignoring the sets with subscript i.

The definition of independence of events can be formulated in terms of independence of σ-algebras. Recall that the σ-algebra $\sigma(\{A\})$ generated by a single event A is $\{\emptyset, \Omega, A, A^c\}$.

Theorem 5.7
Events $A_1, A_2 \ldots, A_n$ are mutually independent if and only if each of the corresponding σ-algebras $\sigma(\{A_1\}), \sigma(\{A_2\}) \ldots, \sigma(\{A_n\})$ are independent.

We define the independence of n random variables via the generated σ-algebras.

Definition 5.8

Random variables $X_1, X_2, \ldots, X_n$ are said to be *mutually independent*, or simply *independent*, if $\sigma(X_1), \sigma(X_2), \ldots, \sigma(X_n)$ are independent.

When there are infinitely many sub-σ-algebras, they are defined to be independent if any finite combination of them is independent.

Definition 5.9

Let $(\Omega, \mathscr{F}, P)$ be a probability space and $(\mathscr{F}_i)_{i=1}^{\infty}$ be a sequence of sub-σ-algebras in $\mathscr{F}$. We say that the σ-algebras $\mathscr{F}_1, \mathscr{F}_2, \mathscr{F}_3, \ldots$ are independent if for any positive integer k, any distinct indices $i_1, i_2, \ldots, i_k \in \mathbb{N}$, and any choices of event $A_j \in \mathscr{F}_{i_j}$, for $j = 1, 2, \ldots, k$, we have

$$P(A_1 \cap A_2 \cap \cdots \cap A_k) = P(A_1)P(A_2) \cdots P(A_k).$$

In terms of independence of σ-algebras, we have a unified definition of independence of infinitely many events and infinitely many random variables.

Definition 5.10

An infinite sequence of events $(A_i)_{i=1}^{\infty}$ in Ω is said to be *independent* if the generated σ-algebras $(\sigma(\{A_i\}))_{i=1}^{\infty}$ are independent.
An infinite sequence of random variables $(X_i)_{i=1}^{\infty}$ in Ω is said to be *independent* if the generated σ-algebras $(\sigma(X_i))_{i=1}^{\infty}$ are independent.

It is important to note that in the definition of independence, it is not necessary to consider the intersection of infinitely many events. Instead, it suffices to consider only finite intersections. For example, consider a sequence of events $A_1, A_2, A_3, \ldots$, and suppose that any finite combination of them is independent. The infinite intersection $\cap_{i=1}^{\infty} A_i$ can be expressed as the limit of a decreasing sequence of finite intersections $\cap_{i=1}^{k} A_i$, for $k \geq 1$. By applying the upper semi-continuity of measure, we can show that the probability $P(\cap_{i=1}^{\infty} A_i)$ is equal to the limit of $P(\cap_{i=1}^{k} A_i)$, as $k \to \infty$. Since the first k events in the sequence are independent, we have

$$P(\cap_{i=1}^{\infty} A_i) = \lim_{k \to \infty} P(\cap_{i=1}^{k} A_i) = \lim_{k \to \infty} \prod_{i=1}^{k} P(A_i) \triangleq \prod_{i=1}^{\infty} P(A_i).$$

Because we are going to frequently investigate independent random variables with the same distribution, we introduce the short-hand notation "i.i.d." to mean that a sequence of random variables is independent and identically distributed.

5.4   Borel–Cantelli Lemmas

We use the notation $P(A_i \ i.o.)$ to represent the probability that event A_i happens for infinitely many indices i. The short-hand notation "i.o." stands for "infinitely often."

Theorem 5.8 (Borel–Cantelli (BC) Lemma 1)
If A_i are events in a probability space $(\Omega, \mathscr{F}, P)$, for $i \geq 1$, satisfying $\sum_{i=1}^{\infty} P(A_i) < \infty$, then

$$P(\limsup_i A_i) = P(A_i \ i.o.) = 0.$$

Proof We can write the limsup of A_i's as an intersection $\limsup_i A_i = \bigcap_{k=1}^{\infty} E_k$, where E_k is the event $E_k \triangleq \bigcup_{j=k}^{\infty} A_j$. Since $E_1 \supseteq E_2 \supseteq E_3 \supseteq \cdots$ is a decreasing sequence of events, by the upper semi-continuity of measure, it suffices to prove that the probability $P(E_k)$ approaches 0 as $k \to \infty$.

Since the probability measure P satisfies the σ-subadditivity property (Definition 3.8), we get

$$P(E_k) = P(\cup_{j=k}^{\infty} A_j) \leq \sum_{j=k}^{\infty} P(A_j).$$

Since $\sum_{j=1}^{\infty} P(A_j)$ is a convergence series, the limit $\lim_{k \to \infty} \sum_{j=k}^{\infty} P(A_j)$ must be zero. This implies $P(E_k) \searrow 0$ as $k \to \infty$. ∎

Example 5.4.1 (Illustration of the First BC Lemma)
Consider a random experiment of tossing a fair coin repeatedly an infinite number of times. Let A_i be the event that there is a run of i heads starting from position i, for $i = 1, 2, 3, \ldots$. Since $\sum_{i=1}^{\infty} P(A_i) = \sum_{i=1}^{\infty} 2^{-i} < \infty$, we have $P(A_i \text{ i.o.}) = 0$.

Note that the events A_i in the previous example are *not* independent, but we can apply the first BC lemma as there is no assumption of independence in this lemma.

In the second Borel–Cantelli lemma, the assumption of independence is essential.

Theorem 5.9 (Borel–Cantelli (BC) Lemma 2)
If $(A_i)_{i=1}^{\infty}$ is a sequence of independent events in a probability space $(\Omega, \mathscr{F}, P)$ such that $\sum_{i=1}^{\infty} P(A_i) = \infty$, then

$$P(\limsup_i A_i) = P(A_i \text{ i.o.}) = 1.$$

Proof Consider the complement of the event $\limsup_i A_i$

$$\left(\limsup_i A_i \right)^c = \left(\bigcap_{k=1}^{\infty} \bigcup_{j=k}^{\infty} A_j \right)^c = \bigcup_{k=1}^{\infty} \bigcap_{j=k}^{\infty} A_j^c.$$

We want to show that $P\left(\cup_{k=1}^{\infty} \cap_{j=k}^{\infty} A_j^c \right) = 0$. It suffices to prove that $P\left(\cap_{j=k}^{\infty} A_j^c \right) = 0$ for all k.

We use the inequality $1 - x \leq e^{-x}$, which holds for all $x \in \mathbb{R}$. For any k, consider an integer m larger than k

$$P\Big(\bigcap_{j=k}^{m} A_j^c\Big) = \prod_{j=k}^{m} P(A_j^c) = \prod_{j=k}^{m}(1 - P(A_j)) \le \prod_{j=k}^{m} e^{-P(A_j)} = e^{-\sum_{j=k}^{m} P(A_j)}.$$

In the first equality above, we use the property that $A_k^c, \ldots, A_m^c$ are independent events (Theorem 5.6). Since $\sum_{i=1}^{\infty} P(A_i)$ is a divergent series, we have $\sum_{j=k}^{m} P(A_j) \to \infty$ as $m \to \infty$. Therefore, using continuity from above, we obtain $P(\cap_{j=k}^{\infty} A_j^c) = 0$. Since it is true for all k, we prove $P(\cup_{k=1}^{\infty} \cap_{j=k}^{\infty} A_j^c) = 0$. ∎

Example 5.4.2 (Monkey Randomly Typing on a Keyboard)

Let $\mathcal{A}$ be an alphabet set of finite size, and let $w = a_1 a_2 \ldots a_n$ be a particular word of length n, where $a_i \in \mathcal{A}$ for all i. If we randomly generate an infinite string $(X_i)_{i=1}^{\infty}$, where X_i's are independent and uniformly chosen from $\mathcal{A}$, then with probability 1, the word w appears at least once in the infinite string in n consecutive positions.

To see this, we can divide the infinite string into infinitely many non-overlapping sub-strings of length n. The first sub-string is from position 1 to n, the second sub-string is from position $n + 1$ to $2n$, and so on. For $k = 1, 2, 3, \ldots$, let A_k be the event that the k-th sub-string is exactly equal to the chosen string w. These events are independent because there is no overlap between the sub-strings. Although the probability $P(A_k)$ is a very small number, it is still strictly positive. Hence, by the second Borel–Cantelli lemma, with probability 1, infinitely many such sub-strings are equal to w.

This example is often presented vividly by taking the word w to be a Shakespeare novel and imagining a monkey randomly typing on a keyboard one character at a time. The Borel–Cantelli lemma states that given infinite time, the monkey will reproduce the entire volume of the Shakespeare novel, with probability 1.

Combining the two Borel–Cantelli lemmas, we have the following zero–one law.

Theorem 5.10 (Borel's Zero–One Law)
Suppose $(A_i)_{i=1}^{\infty}$ is a sequence of independent events in a probability space $(\Omega, \mathcal{F}, P)$. Then

$$P(A_i \ i.o.) = \begin{cases} 1 & \text{if } \sum_{i=1}^{\infty} P(A_i) = \infty \\ 0 & \text{if } \sum_{i=1}^{\infty} P(A_i) < \infty. \end{cases}$$

Example 5.4.3 (Infinite Sequence of Independent Random Bits)

Let s be a positive real number. We generate an infinite sequence of independent bits as follows. For $n = 1, 2, 3, \ldots$, the n-th bit is equal to 1 with probability $1/n^s$ and is equal to 0 with probability $1 - 1/n^s$. If $\sum_{n=1}^{\infty} 1/n^s$ is finite, then with probability 1, there are finitely many 1's in the generated bit stream. On the other hand, if $\sum_{n=1}^{\infty} 1/n^s$ diverges, then with probability 1 there are infinitely many 1's.

This example exhibits a phase transition at $s = 1$. We have infinitely many 1's almost surely when $s \le 1$, but finitely many 1's almost surely when $s > 1$.

5.5   A Model for a Sequence of Independent Random Variables

In this section we show how to construct a countably infinite sequence of independent random variables by drawing a random number uniformly in the unit interval $[0, 1]$. The uniform distribution is a fundamental building block in the probability theory, as it contains enough randomness to generate an entire sequence of independent random variables. With this probability model, we prove the Kolmogorov zero–one law.

For any real number u between 0 and 1, we let $B_i(u)$ be the i-th bit in the binary expansion of u, i.e.,

$$u = \sum_{i=1}^{\infty} B_i(u) 2^{-i}.$$

When a number has more than one binary expansion, we choose the one with infinitely many trailing zeros. For any real number u, we obtain an infinite binary sequence $(B_i(u))_{i \geq 1}$. Note that the sequences with infinitely many trailing ones are not represented in this probability model. However, these missing sequences form a set of probability 0 and are therefore negligible.

The first bit $B_1(u)$ is equal to 0 if $u \in A_1 = [0, 1/2)$ and 1 if $u \in [1/2, 1)$. The second bit $B_2(u)$ equals 0 if $u \in A_2 = [0, 1/4) \cup [1/2, 3/4)$ and 1 if u is in the complement of A_2 in $[0, 1)$. In general, the first n bits of u are $x_1, x_2, \ldots, x_n$, where x_i are binary digits in $\{0, 1\}$, when

$$\sum_{i=1}^{n} x_i 2^{-i} \leq u < \frac{1}{2^n} + \sum_{i=1}^{n} x_i 2^{-i}. \tag{5.5}$$

The mapping from a real number u to the associated sequence $(B_i(u))_{i=1}^{\infty}$ described in the previous paragraph is a deterministic procedure. Specifically, $B_i(u)$ is the i-th bit in the binary expansion of u. We now construct a Lebesgue–Stieltjes measure on $[0, 1]$ using the Stieltjes measure function $F(x) = x$, which is the identity function, for $0 \leq x \leq 1$. This measure, denoted by P, defines a probability space $([0, 1], \mathscr{B}([0, 1]), P)$.

The i-th bit $B_i(u)$ becomes a random variable under the probability measure P. For $x_1, x_2, \ldots, x_n \in \{0, 1\}$, the probability that the bit sequence $B_i = x_i$ for $i = 1, 2, \ldots, n$ is

$$P(B_1 = x_1, \ B_2 = x_2, \ldots, B_n = x_n) = \frac{1}{2^n} \tag{5.6}$$

by (5.5). This process generates a sequence of independent Bernoulli random variables, each with probability 1/2 of being 0 or 1, and the joint distribution of these random variables is uniform on the set of all binary sequences with length n.

We divide the bits into two streams. The first stream consists of the bits $B_i(u)$ with even indices i and the second with odd indices i,

$$S_e(u) \triangleq (B_2(u), B_4(u), B_6(u), \ldots), \qquad S_o(u) \triangleq (B_1(u), B_3(u), B_5(u), \ldots).$$

We can define two new random variables $X_e(u)$ and $X_o(u)$ by

$$X_e(u) \triangleq \sum_{i=1}^{\infty} B_{2i}(u) 2^{-i}, \qquad X_o(u) \triangleq \sum_{i=1}^{\infty} B_{2i-1}(u) 2^{-i}.$$

The two random variables X_e and X_o are independent and uniformly distributed on $[0, 1)$. They are independent because the bits with even indices and the bits with odd indices are independent. We may call this process "splitting."

We let U_1 to be the random variable X_e. We then split the bits stream S_o into odd and even parts and use one part to define a uniform random variable U_2. This process can be repeated recursively to produce an infinite sequence of independent uniform random variables.

We have thus constructed an infinite sequence $(U_i(u))_{i \geq 1}$ of i.i.d. uniform random variables from a uniformly distributed random variable. By using the inverse transform method (see Problem 5.7), we can transform this sequence to an infinite sequence of independent random variables with arbitrary distribution. For example, if we want a sequence of independent but biased coin tosses, we can proceed as follows. For $i = 1, 2, \ldots$, let Y_i be a binary random variable defined by $Y_i(u) = \mathbf{1}_{\{U_i(u) < p_i\}}$, where $p_1, p_2, p_3, \ldots$ are given constants. The sequence $(Y_i(u))_{i=1}^{\infty}$ represents a sequence of biased but independent coin tosses. It is important to note that once a sample point u between 0 and 1 is selected, the whole sequence $(Y_i(u))_{i=1}^{\infty}$ is determined. This construction procedure provides an explicit probability model for an infinite sequence of independent random variables.

Kolmogorov's zero–one law states that the conclusion in the Borel zero–one law is a common phenomenon. We will use the notation $\sigma(X_1, X_2, X_3, \ldots)$ to denote the smallest σ-algebra under which the random variables X_i's are all measurable.

Theorem 5.11   (Kolmogorov's Zero–One Law)
Let $(X_n)_{n=1}^{\infty}$ be a sequence of random variables, and define the tail σ-algebra by $\mathscr{T} \triangleq \cap_{n=1}^{\infty} \sigma(\{X_i : i \geq n\})$. If the random variables X_n, for $n \geq 1$, are mutually independent, then any event in the tail σ-algebra has probability equal to either 0 or 1.

Proof   The idea of proof is to show that the tail σ-algebra $\mathscr{T}$ is independent of itself. Let $\mathscr{G}_n$ denote the σ-algebra generated by the first n random variables $X_1, \ldots, X_n$ in the sequence. The σ-algebra $\mathscr{G}_n$ is independent of the tail σ-algebra $\mathscr{T}$, because any event in $\mathscr{T}$ is in $\sigma(\{X_i : i > n\})$, which is independent with $\mathscr{G}_n$.

Let $\mathcal{G}$ denote the union $\cup_{n=1}^{\infty} \mathcal{G}_n$. One can show that $\mathcal{G}$ is a field, and this field generates $\sigma(X_1, X_2, \ldots)$. Since $\mathcal{G}_n$ and $\mathcal{T}$ are independent, any set in $\mathcal{G}$ and any set in $\mathcal{T}$ are independent. Using the fact that $\mathcal{G}$ is a π-system and an argument similar to the proof of Theorem 5.5, we can show that $\sigma(X_1, X_2, \ldots)$ is independent of the tail σ-algebra $\mathcal{T}$. Because $\mathcal{T}$ is contained in $\sigma(X_1, X_2, \ldots)$, any set A in $\mathcal{T}$ is in $\sigma(X_1, X_2, \ldots)$ and hence is independent of itself. In particular, we have $P(A) = P(A \cap A) = P(A)P(A)$. Therefore, we conclude that $P(A)$ is equal to either 0 or 1. ∎

Problems

5.1. Let X and Y be continuous random variables with joint pdf $f_{XY}(x, y)$. Prove that if $f_{XY}(x, y)$ factorizes as a product $g(x)h(y)$ of a function $g(x)$ of x and a function $h(y)$ of y, then X and Y are independent.

5.2. Prove that in any probability space $(\Omega, \mathcal{F}, P)$, the empty set is independent of any event in the σ-algebra $\mathcal{F}$.

5.3. Suppose A_1, A_2, A_3, and A_4 are four events in a probability space, and they satisfy

$$P(A_1 \cap A_2 \cap A_3 \cap A_4) = P(A_1)P(A_2)P(A_3)P(A_4).$$

Can we deduce that A_1, A_2, A_3, and A_4 are mutually independent?

5.4. By giving a counter-example, show that the independence assumption in the second Borel–Cantelli lemma cannot be relaxed.

5.5. Show that the independence of two random variables depends on the choice of probability measure. Give an example of measurable space $(\Omega, \mathcal{F})$, two random variables $X(\omega)$ and $Y(\omega)$, and two probability measures P and Q, such that X and Y are independent with probability measure P but are dependent with probability measure Q.

5.6. Suppose we perform a sequence of infinitely many experiments. For $k = 1, 2, 3, \ldots$, the outcome of the k-th experiment is successful with probability $k^{-\alpha}$, where α is a constant strictly between 0 and 1, but is not successful with probability $1 - k^{-\alpha}$. Assume that the experiments are independent of each other. Calculate the probability that we have n consecutive successful experimental outcomes infinitely often, where n is a positive integer.

5.7 (Inverse Transform Method). Let $U_1, U_2, \ldots$ be a sequence of independent random variables uniformly distributed between 0 and 1. Given a Stieltjes measure function $F(x)$ that is continuous and strictly monotonically increasing, such that the inverse function F^{-1} is well-defined, show that $F^{-1}(U_i)$, for $i \geq 1$, is a sequence of i.i.d. random variable with cdf $F(x)$. (For Stieltjes measure function that is not continuous or not strictly monotonic, see the proof of Theorem 10.8 or [4, Theorem 1.2.2].)

5.8. Let Z be a standard normal random variable, and let C be the outcome of a fair coin that is independent of Z. We define the random point (X, Y) as follows: If the outcome of the coin is head, then $(X, Y) = (Z, 0)$, and if the outcome is tails, then $(X, Y) = (0, Z)$.

(i) Show that X and Y are identically distributed but not independent.
(ii) Derive the joint cumulative distribution function $F_{XY}(x, y)$ of X and Y.
(iii) Derive the joint cumulative distribution function $F_{XZ}(x, z)$ of X and Z.

5.9. Let $(\Omega, \mathcal{F}, P)$ denote a probability space and let $\mathcal{G}$ and $\mathcal{H}$ be two independent sub-σ-algebras of $\mathcal{F}$. Prove that any random variable X that is measurable with respect to both $\mathcal{G}$ and $\mathcal{H}$ is equal to a constant with probability 1, i.e., there exists a constant c such that $P(X = c) = 1$.

Lebesgue Integral and Mathematical Expectation

6

Consider a sequence of random variables $X_n(\omega)$ that converge pointwise to a function $X(\omega)$. In Chap. 4, we showed that $X(\omega)$ is measurable. In many cases, it is of interest to know whether the limit of the expected values $E[X_n]$ will converge to the expected value of the limit $E[X]$. However, we cannot always exchange the order of limit and expectation, as there exist examples that give different results depending on the order of operations. Therefore, it is important to find conditions under which such an exchange is justified.

For Riemann integration, there is a useful condition called "uniform convergence." If we can show that the sequence of functions are Riemann integrable and that the convergence is uniform, then the limit function is Riemann integrable and we can exchange the order of limit and integration. However, the notion of uniform convergence requires a distance function defined on the domain of the functions, making it a metric space. In the probability theory, the sample space may be an abstract space, and no such distance function is available. Therefore, we need a way to perform integration on a general probability space, and this is accomplished by Lebesgue integration.

Lebesgue's approach to integration differs from Riemann's by dividing the range of functions into small intervals. The derivation of Lebesgue integral proceeds in several stages. First it is defined for the class of simple functions and then extended to nonnegative functions and, more generally, to real-valued functions. From there, the range is further extended to include the extended real numbers and, separately, the complex numbers. The relationship between these stages of the Lebesgue integral is illustrated below.

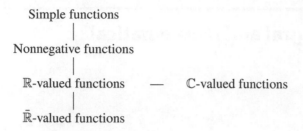

6.1   Simple Functions

In this chapter we will work with a fixed measure space $(\Omega, \mathscr{F}, \mu)$, where μ may be the Lebesgue measure, the counting measure, or a probability measure.

Definition 6.1

Let $(\Omega, \mathscr{F}, \mu)$ be a measure space. A function $X : \Omega \to \bar{\mathbb{R}}$ is called a *simple function* if its range is finite and it is $\mathscr{F}$-measurable.

Suppose X takes on n distinct values, denoted by $a_1, a_2, \ldots, a_n$, for some integer n. Since X is a function with values in $\bar{\mathbb{R}}$, one of the a_i's may be ∞ or $-\infty$. For $i = 1, 2, \ldots, n$, let A_i be the pre-image $X^{-1}(\{a_i\}) \triangleq \{\omega : X(\omega) = a_i\}$. Because a simple function is $\mathscr{F}$-measurable by definition, the sets A_i's are also $\mathscr{F}$-measurable and form a partition of the sample space Ω. We can write $X(\omega)$ as a linear combination of indicator functions

$$X(\omega) = \sum_{i=1}^{n} a_i \mathbf{1}_{A_i}(\omega). \tag{6.1}$$

Definition 6.2

The *Lebesgue integral* of a simple function X expressed as in (6.1) is defined by

$$\int X \, d\mu \triangleq \sum_{i=1}^{n} a_i \mu(A_i), \tag{6.2}$$

unless the summation contains both ∞ and $-\infty$, in which case the integral is not defined.

When working with infinity in (6.2), we adopt the following conventions:

▶ **Convention with ∞ and 0**

- A term in (6.2) equals ∞ if (i) $a_i > 0$ and $\mu(A_i) = \infty$, or $a_i = \infty$ and $\mu(A_i) > 0$.
- If all terms in (6.2) are either finite or ∞, the integral equals a real number or ∞.
- If $a_i < 0$ and $\mu(A_i) = \infty$, then $a_i \mu(A_i)$ equals $-\infty$. The integral is equal to $-\infty$ if all the other terms are either finite or $-\infty$.
- If $a_i = 0$ or $\mu(A_i) = 0$, the product $a_i \mu(A_i)$ is defined as 0 by convention, and it makes no contribution to the integral.

▶ **Notation for Lebesgue Integral** We will define the Lebesgue integral for a more general class of functions later. To emphasize that ω is the variable of integration, we can write

$$\int X(\omega)\,d\mu(\omega) \quad \text{or} \quad \int X(x)\,d\mu(x) \quad \text{or} \quad \int X(\omega)\,\mu(d\omega) \quad \text{or} \quad \int X(x)\,\mu(dx).$$

If μ is understood from the context, we will use the shorthand notation $\int X$.

Example 6.1.1 (Lebesgue Integral for Finite Sample Space)
When Ω is a finite set and μ is the counting measure, the Lebesgue integral reduces to a finite sum. If Ω is a finite set and μ is a general measure on Ω, then the Lebesgue integral of a simple function is a finite weighted sum.

Example 6.1.2 (Lebesgue Integral of a Single Indicator Function)
Suppose B is a measurable set. By Theorem 4.1, the indicator function $\mathbf{1}_B$ is measurable. It takes two values and can be expressed as

$$\mathbf{1}_B(\omega) = 1 \cdot \mathbf{1}_B(\omega) + 0 \cdot \mathbf{1}_{B^c}(\omega).$$

The Lebesgue integral $\int \mathbf{1}_B\,d\mu$ equals $1 \cdot \mu(B) + 0 \cdot \mu(B^c) = \mu(B)$, where the second term $0 \cdot \mu(B^c)$ is 0 by the convention of $0 \cdot x = 0$.

Theorem 6.1
Let X and Y be simple functions, and let $\alpha \in \mathbb{R}$ be a constant. Then,

(a) $\int \alpha X\,d\mu = \alpha \int X\,d\mu$.
(b) $\int X + Y\,d\mu = \int X\,d\mu + \int Y\,d\mu$.
(c) if $X \leq Y$, then $\int X\,d\mu \leq \int Y\,d\mu$.

The two properties in parts (a) and (b) are called the *linearity property* of integral. The property in part (c) is called the *monotonic property*.

Proof

(a) Suppose X is expressed as $X = \sum_{i=1}^{m} a_i \mathbf{1}_{A_i}$, where $A_i = X^{-1}(\{a_i\})$ for all i, and the sets A_i's form a partition of Ω. If $\alpha = 0$, then both sides of (a) are zero. When $\alpha \neq 0$, the function αX takes m distinct values and can be written as $\sum_{i=1}^{m} (\alpha a_i) \mathbf{1}_{A_i}$. Hence

$$\int aX = \sum_{i=1}^{m} (\alpha a_i) \mu(A_i) = \alpha \sum_{i=1}^{m} a_i \mu(A_i) = \alpha \int X.$$

(b) We continue with the notation in part (a) and write $Y = \sum_{j=1}^{n} b_j \mathbf{1}_{B_j}$, where $B_j = Y^{-1}(\{b_j\})$ and the sets B_j's form a partition of Ω. The function $X + Y$ is measurable because both X and Y are measurable. The number of values that $X + Y$ can take is no more than the number of values of X times the number of values of Y and hence is finite. This verifies that $X + Y$ is a simple function.

We can consider the refined partition $\{A_i \cap B_j : i = 1, \ldots, m, j = 1, \ldots, n\}$ of Ω and write $X + Y$ as $\sum_{i=1}^{m} \sum_{j=1}^{n} (a_i + b_j) \mathbf{1}_{A_i \cap B_j}$. The set $A_i \cap B_j$ may be empty for some i and j, but it will not affect the integral, as $A_i \cap B_j$ has zero measure in this case and makes zero contribution to the integral. The Lebesgue integral of $X + Y$ is equal to

$$\sum_{i=1}^{m} \sum_{j=1}^{n} (a_i + b_j) \mu(A_i \cap B_j) = \sum_{i=1}^{m} \sum_{j=1}^{n} a_i \mu(A_i \cap B_j) + \sum_{i=1}^{m} \sum_{j=1}^{n} b_j \mu(A_i \cap B_j)$$

$$= \sum_{i=1}^{m} a_i \mu(A_i) + \sum_{j=1}^{n} b_j \mu(B_j),$$

which is the same as $\int X \, d\mu + \int Y \, d\mu$.

(c) We first prove the inequality with $X = 0$ and $Y \geq 0$. When $Y \geq 0$, then $b_j \geq 0$ for all j, and thus

$$\int Y \, d\mu = \sum_{j=1}^{n} b_j \mu(B_j) \geq 0.$$

In general, when $X \leq Y$, then $Y - X \geq 0$, and $0 \leq \int Y - X \, d\mu = \int Y - \int X$ using linearity property of integral in (b). We then get $\int X \leq \int Y$. ∎

Theorem 6.2
Suppose

$$X = \sum_{j=1}^{n} b_j \mathbf{1}_{B_j},$$

where B_j's are measurable sets (not necessarily mutually disjoint). Then X is a simple function, and

$$\int X \, d\mu = \sum_{j=1}^{n} b_j \mu(B_j).$$

Proof By the linearity of Lebesgue integral for simple functions,

$$\int X \, d\mu = \sum_{j=1}^{n} b_j \int \mathbf{1}_{B_j} \, d\mu = \sum_{j=1}^{n} b_j \mu(B_j).$$

In the second equality in the above line, we have applied Example 6.1.2 to evaluate the integral of indicator functions. ∎

6.2 Lebesgue Integral of Nonnegative Functions

In this section we define the Lebesgue integral for measurable function whose values are either nonnegative real numbers or ∞. Unlike the Riemann–Stieltjes integral, we partition the *range* of X into subintervals and take limit as the width of the subintervals tends to zero.

To define the Lebesgue integral of a nonnegative function, we first approximate it from below by simple functions and then take the supremum of the integrals of these simple functions.

Definition 6.3

For a nonnegative measurable function X, we define the *Lebesgue integral* of X by

$$\int X \, d\mu \triangleq \sup \left\{ \int f \, d\mu : f \text{ is simple}, 0 \le f \le X \right\}.$$

If we define the set

$$S(X) \triangleq \{f : f \text{ is simple, } 0 \le f \le X\}, \tag{6.3}$$

then the Lebesgue integral can be expressed as

$$\int X \, d\mu = \sup \left\{ \int f \, d\mu : f \in S(X) \right\}.$$

If the set $\{\int f \, d\mu : f \in S(X)\}$ is unbounded, then $\int X \, d\mu = \infty$. Furthermore, if $X(\omega)$ is equal to ∞ for ω in a set A with positive measure, then the integral $\int X \, d\mu$ is also equal to ∞.

We note that the set $S(X)$ in (6.3) is not empty. One way to approximate a nonnegative measurable function X by simple functions is as follows. For any positive integer n, we define a simple function $X^{(n)}$ by

$$X^{(n)}(\omega) = \begin{cases} n & \text{if } n < X(\omega), \\ \frac{k}{2^n} & \text{if } \frac{k}{2^n} < X(\omega) \le \frac{k+1}{2^n} \text{ for some } k = 0, 1, 2, \ldots, n2^n - 1. \end{cases} \tag{6.4}$$

By construction, the function $X^{(n)}$ has finite range and we have $X^{(n)}(\omega) \le X(\omega)$ for all ω. We also note that for each fixed ω, the sequence $(X^{(n)}(\omega))_{n \ge 1}$ converges to $X(\omega)$ from below as $n \to \infty$.

▶ **Backward Compatibility** At this point, we have two apparently different definitions of integral for nonnegative simple functions. However, Definitions 6.2 and 6.3 are compatible with each other when X is a simple nonnegative function. Specifically, suppose X is a measurable function expressed as

$$X(\omega) = \sum_{i=1}^{k} c_i \mathbf{1}_{A_i}(\omega),$$

where $c_i \ge 0$ for all i, and the sets A_i's are measurable and mutually disjoint. By Definition 6.2, the integral of X is given by $\sum_{i=1}^{k} c_i \mu(A_i)$. It can be shown that this value is indeed the maximum of $\{\int f \, d\mu : f \in S(X)\}$.

The monotonic property is a direct consequence of the definition.

Theorem 6.3 (Monotonicity)
Suppose X and Y are nonnegative measurable functions, and $X \le Y$. Then $\int X \, d\mu \le \int Y \, d\mu$.

Proof By the definition of Lebesgue integral for nonnegative functions, we have

$$\int X \, d\mu \triangleq \sup \left\{ \int f \, d\mu : f \in S(X) \right\}$$

$$\int Y \, d\mu \triangleq \sup \left\{ \int g \, d\mu : g \in S(Y) \right\}.$$

Since $X \leq Y$, we have $S(X) \subseteq S(Y)$. Therefore, for any $f \in S(X)$, we have $f \in S(Y)$, and so

$$\int f \, d\mu \leq \sup \left\{ \int g \, d\mu : g \in S(Y) \right\}.$$

Taking the supremum over $S(X)$, we get $\int X \, d\mu \leq \int Y \, d\mu$.   ∎

The following theorem summarizes the behavior of integrals for non-decreasing sequences of functions, which is a fundamental result in measure theory and probability theory. We say that a sequence of functions $X_1, X_2, X_3, \ldots$ is non-decreasing if they are non-decreasing pointwise, that is, for each ω in Ω, we have

$$X_1(\omega) \leq X_2(\omega) \leq X_3(\omega) \leq \cdots,$$

and we simply write it as $X_1 \leq X_2 \leq X_3 \leq \cdots$.

***

Theorem 6.4   (Monotone Convergence Theorem (MCT))
If

$$0 \leq X_1 \leq X_2 \leq X_3 \leq \cdots$$

is a sequence of non-decreasing, nonnegative measurable functions converging to X from below, then

$$\int X_n \, d\mu \to \int X \, d\mu$$

as $n \to \infty$.

***

We note that the pointwise limit of $(X_n)_{n=1}^{\infty}$ in Theorem 6.4 always exists. We set $X(\omega) = \infty$ when $\lim_{n \to \infty} X_n(\omega) = \infty$. We use the notation $X_n \nearrow X$ to mean that $(X_n)_{n \geq 1}$ is a non-decreasing sequence converging pointwise to X.

Proof By monotonicity, we have $\int X_n \leq \int X$ for all n. Therefore, $\sup_n \int X_n \leq \int X$. To complete the proof, we need to show that $\sup_n \int X_n \geq \int X$.

Suppose g is a simple function in $S(X)$, so that $g(\omega) \leq X(\omega)$ for all ω. We write $g(\omega)$ as a finite summation $g(\omega) = \sum_{i=1}^{k} a_i \mathbf{1}_{A_i}(\omega)$, where $a_i \geq 0$ for $i = 1, 2, \ldots, k$, and the sets A_i's are measurable and mutually disjoint.

Fix $\epsilon > 0$. For any $n \geq 1$, let

$$B_n^\epsilon \triangleq \{\omega \in \Omega : X_n(\omega) \geq (1 - \epsilon)g(\omega)\}.$$

The set B_n^ϵ is measurable because both X_n and g are measurable functions.

Since $X_n \nearrow X$ by assumption and $g \leq X$, we have $B_n^\epsilon \nearrow \Omega$ as $n \to \infty$. For each n, we can restrict the integral to the set B_n^ϵ and lower bound $X_n(\omega)$ by $(1 - \epsilon)g(\omega)$. Hence, using the monotonicity of Lebesgue integral, we obtain

$$\int X_n \, d\mu \geq \int \mathbf{1}_{B_n^\epsilon} X_n \, d\mu \geq \int \mathbf{1}_{B_n^\epsilon}(\omega)(1 - \epsilon)g(\omega) d\mu(\omega)$$

$$= (1 - \epsilon) \int \mathbf{1}_{B_n^\epsilon} \sum_{i=1}^{k} a_i \mathbf{1}_{A_i} \, d\mu = (1 - \epsilon) \sum_{i=1}^{k} a_i \mu(B_n^\epsilon \cap A_i).$$

Since $B_n^\epsilon \nearrow \Omega$, we can take limits on both sides as $n \to \infty$. This yields

$$\sup_n \int X_n \, d\mu \geq (1 - \epsilon) \sum_{i=1}^{k} a_i \mu(A_i).$$

Because ϵ can be arbitrarily small, we get $\sup_n \int X_n \, d\mu \geq \int g \, d\mu$. Since the above inequality holds for any $g \in S(X)$, we obtain $\sup_n \int X_n \geq \int X$. ∎

Using the monotone convergence theorem, we can now prove the linearity property.

> **Theorem 6.5 (Linear Property)**
> *Let X and Y be nonnegative and measurable functions and let α be a nonnegative constant. Then*
>
> $$\int X + Y \, d\mu = \int X \, d\mu + \int Y \, d\mu,$$
>
> $$\int \alpha X \, d\mu = \alpha \int X \, d\mu.$$

The equalities in this theorem are interpreted as follows: If the one side of an equality in Theorem 6.5 is infinity, then the other side is also infinity.

Proof Let X_n's and Y_n's be simple nonnegative functions such that $X_n \nearrow X$ and $Y_n \nearrow Y$. Such sequences of simple functions always exist using the approximation in (6.4).

For each n, the sum $X_n + Y_n$ is simple nonnegative functions converging to $X + Y$ from below. By applying the monotone convergence theorem, we obtain

$$\int X + Y \, d\mu = \int \lim_{n \to \infty} (X_n + Y_n) \, d\mu$$

$$= \lim_{n \to \infty} \int (X_n + Y_n) \, d\mu$$

$$= \int X \, d\mu + \int Y \, d\mu.$$

To prove the second equality, we approximate αX by $(\alpha X_n)_{n=1}^{\infty}$, which is non-decreasing and converging to αX from below. The proof is similar to the first part. ∎

6.3   Lebesgue Integral of Real-Valued and Complex-Valued Functions

Definition 6.4

For real-valued measurable function X, we let $X^+ \triangleq \max(X, 0)$ and $X^- \triangleq -\min(X, 0)$ be the *positive part* and *negative part* of X, respectively.

Both the positive and negative parts of X are measurable because the functions $\max(x, 0)$ and $-\min(x, 0)$ are continuous (Theorem 4.5). Moreover, we have

$$X = X^+ - X^- \quad \text{and} \quad |X| = X^+ + X^-.$$

Using these concepts, we can define the Lebesgue integral of a real-valued function by reducing the definition to the nonnegative case.

Definition 6.5 (Real Lebesgue Integral)

We define the *Lebesgue integral* of a measurable $\bar{\mathbb{R}}$-valued function X on a measure space $(\Omega, \mathscr{F}, \mu)$ by

$$\int X \, d\mu \triangleq \int X^+ \, d\mu - \int X^- \, d\mu.$$

It is well-defined unless we have $\infty - \infty$ on the right-hand side. When $\int X^+$ and $\int X^-$ are both finite, we say that X is μ-*integrable* and write $X \in L^1(\mu)$.

To simplify the notation, we will write "X is integrable" if the measure μ is known from the context. If $\int X\,d\mu = \infty$ or $\int X\,d\mu = -\infty$, the function X is not integrable, but the integral is still considered well-defined. An example of a function with an undefined integral is demonstrated below.

Example 6.3.1 (An Example of Undefined Integral)

Suppose $\Omega = \mathbb{Z}$ is the set of integers equipped with discrete σ-field $\mathscr{F} = \mathscr{P}(\mathbb{Z})$. Define a probability measure on $\mathbb{Z}$ by $P(\{\omega\}) = p_\omega$, where $p_0 = 0$, and $p_k = p_{-k} = c/k^2$ for $k = 1, 2, 3, \ldots$. The constant c is chosen such that $\sum_{k \in \mathbb{Z}} p_k = 1$. Consider the function $X(\omega) = \omega$. The integral $\int \omega\,dP(\omega)$ is not defined because the positive and negative parts are infinite

$$\int X^+\,dP = \sum_{k=1}^{\infty} k\frac{c}{k^2} = c\sum_{k=1}^{\infty}\frac{1}{k} = \infty,$$

$$\int X^-\,dP = \sum_{k=1}^{\infty} k\frac{c}{k^2} = \infty.$$

The distribution in the example is a discrete analogue of the Cauchy distribution. In probability theory, the Cauchy distribution is defined by the pdf

$$f(x) = (\pi(x^2 + 1))^{-1}$$

for $x \in \mathbb{R}$. The calculation of its expectation requires that we first compute the positive and negative Lebesgue integrals of $xf(x)$, respectively. However, in the case of Cauchy distribution, both the positive and negative Lebesgue integrals are infinite. Hence, even though the pdf is symmetric around the origin, the expectation is undefined.

▶ **Convention on Integration over a Set** In the Lebesgue integration, integrating a measurable function over a measurable subset E of Ω is equivalent to multiplying the integrand by the indicator function $\mathbf{1}_E(\omega)$ that takes the value 1 on E and 0 elsewhere. We denote this operation by

$$\int_E X\,dP \triangleq \int X\mathbf{1}_E\,dP.$$

If we want to emphasize that the integration is over the entire sample space, we use the notation $\int_\Omega X\,dP$.

In the field of complex numbers, we do not have a natural ordering, and there is no $-\infty$. We define the Lebesgue integral of a complex-valued function in terms of the Lebesgue integral of its real and imaginary parts. Specifically, motivated by the linearity of integral, we add the Lebesgue integrals of the real and imaginary parts to obtain the Lebesgue integral of the complex-valued function. To avoid confusion, we generally assume that the complex-valued function to be integrated does not take ∞ as its value.

Definition 6.6 (Complex Lebesgue Integral)

Suppose $Z(\omega) = X(\omega) + iY(\omega)$, where $X(\omega)$ and $Y(\omega)$ are the real and imaginary parts of $Z(\omega)$, respectively. If both X and Y are integrable, then we say that Z is *integrable* and define the Lebesgue integral of Z by

$$\int Z \, d\mu \triangleq \int X \, d\mu + i \int Y \, d\mu$$

and write $Z \in L^1(\mu)$. The integral of Z is not defined if X or Y is not integrable.

The integral of complex-valued function will be used later in the study of a characteristic function. We give a simple example below.

Example 6.3.2 (An Example of Complex Integral)
Take $\Omega = [0, 1]$ as the sample space and μ be the Lebesgue measure on $[0, 1]$. Assume for the time being that the integral with respect to the Lebesgue measure can be computed by Riemann integral. (This will be proved in the next chapter.)
Let t be a real number. We are interested in computing the integral

$$\phi(t) = \int_\Omega e^{ixt} \, d\mu(x)$$

as a function of t. By using Euler's formula, we can write e^{ixt} as $\cos(xt) + i\sin(xt)$ and compute the complex integral as

$$\phi(t) = \int_\Omega e^{ixt} \, d\mu(x) \triangleq \int_0^1 \cos(xt) \, dx + i \int_0^1 \sin(xt) \, dx.$$

When t is nonzero, it is equal to

$$\left[\frac{\sin(xt)}{t}\right]_0^1 + i\left[-\frac{\cos(xt)}{t}\right]_0^1 = \frac{\sin(t) - i\cos(t) + i}{t} = \frac{i - ie^{it}}{t}.$$

The next theorem provides a useful criterion for determining whether a measurable function is integrable.

Theorem 6.6
An $\bar{\mathbb{R}}$-valued or $\mathbb{C}$-valued measurable function X is integrable if and only if $\int |X| \, d\mu$ is finite.

Proof We first consider an $\bar{\mathbb{R}}$-valued function X. If X is integrable, then by definition $\int X^+$ and $\int X^-$ are finite. This implies $\int |X| = \int X^+ + \int X^-$ is finite (by the linear property of Lebesgue integral for nonnegative function). Conversely, suppose $\int |X|$ is finite. Since $X^+ \leq |X|$ and $X^- \leq |X|$, by the monotonic property in Theorem 6.3, both $\int X^+$ and $\int X^-$ are finite, and hence X is integrable.

Let $Z(\omega) = X(\omega) + iY(\omega)$ denote a complex-valued function. Suppose Z is integrable, i.e., both its real and imaginary parts are integrable. Using the triangle inequality of complex numbers $|x + iy| \le |x| + |y|$ and the monotonic property of real integral, we obtain

$$\int |Z| = \int |X + iY| \le \int |X| + |Y| = \int |X| + \int |Y| < \infty.$$

Conversely, suppose $\int |Z|$ is finite. Since $|X(\omega)| \le |Z(\omega)|$ for all $\omega \in \Omega$, we obtain $\int |X| < \infty$. Similarly, we have $\int |Y| < \infty$. This proves that Z is integrable. ∎

▶ **About Conditional Integrable Function in Riemann Integration** In the theory of Lebesgue integral, there is no such thing as "conditional convergence" or "conditionally integrable." The previous theorem says that integrable function is the same as "absolutely integrable function." This means that if a function is not absolutely integrable, then it is not integrable in the Lebesgue sense, regardless of any "conditional" convergence or integrability properties it may have. This is in contrast to the theory of Riemann integration, where a function can be conditionally convergent or integrable, even if its absolute value is not integrable. As a result, there are functions that are integrable in the Riemann sense but not in the Lebesgue sense.

We now establish the linearity of Lebesgue integral in general.

Theorem 6.7
If X and Y are an $\bar{\mathbb{R}}$-valued or $\mathbb{C}$-valued integrable function and α is a constant, then $X + Y$ and αX are integrable, and

$$\int X + Y \, d\mu = \int X \, d\mu + \int Y \, d\mu,$$

$$\int \alpha X \, d\mu = \alpha \int X \, d\mu.$$

The equalities in this theorem are interpreted as follows: If the one side is well-defined, then the other side is well-defined and the equality holds. If the one side is not defined, then the other side is also not defined. In the case of an $\bar{\mathbb{R}}$-valued function, when the one side of the equation is infinity, then the other side is also infinity.

Proof Suppose X and Y are measurable functions with values in $\bar{\mathbb{R}}$. For all $\omega \in \Omega$, we have $|X(\omega) + Y(\omega)| \leq |X(\omega)| + |Y(\omega)|$. Therefore, by the monotonic property,

$$\int |X + Y| \leq \int |X| + \int |Y|.$$

Since $\int |X|$ and $\int |Y|$ are both finite by the assumption and Theorem 6.6, the integral $\int |X + Y|$ is finite.

For any real number α, the function αX is integrable because

$$\int |\alpha X| = \int |\alpha| |X| = |\alpha| \int |X| < \infty.$$

The proof integrability in the complex case is the similar. We use the triangle inequality $|z_1 + z_2| \leq |z_1| + |z_2|$ for complex numbers z_1 and z_2, and the equality $|\alpha z| = |\alpha||z|$, which holds for constants α and complex numbers z.

For linearity, we first consider the case of $\bar{\mathbb{R}}$-valued measurable functions X and Y. Denote the positive and negative parts of X by X^+ and X^- and the positive and negative parts of Y by Y^+ and Y^-. We write $X + Y$ in two ways:

$$X + Y = (X + Y)^+ - (X + Y)^-$$
$$X + Y = X^+ - X^- + Y^+ - Y^-.$$

Eliminating $X + Y$ from the above two identities, we obtain

$$(X + Y)^+ + X^- + Y^- = (X + Y)^- + X^+ + Y^+.$$

Both the left- and right-hand sides are sum of nonnegative functions and hence are also nonnegative. By the linearity of Lebesgue integral for nonnegative functions,

$$\int (X + Y)^+ + \int X^- + \int Y^- = \int (X + Y)^- + \int X^+ + \int Y^+$$
$$\int (X + Y)^+ - \int (X + Y)^- = \int X^+ - \int X^- + \int Y^+ - \int Y^-$$
$$\int X + Y = \int X + \int Y.$$

When α is a positive real number, $(\alpha X)^+ = \alpha (X)^+$ and $(\alpha X)^- = \alpha (X)^-$. Thus

$$\int \alpha X \triangleq \int (\alpha X)^+ - \int (\alpha X)^- = \alpha \left(\int X^+ - \int X^- \right) = \alpha \int X.$$

When $\alpha = -1$, from $(-X)^+ = X^-$ and $(-X)^- = X^+$, we can deduce

$$\int (-X)\, d\mu \triangleq \int (-X)^+ - \int (-X)^- = \int X^- - \int X^+ = -\int X.$$

This finish the proof of the linear property for $\bar{\mathbb{R}}$-valued functions.

Now suppose that X and Y are complex-valued measurable functions. The first equality $\int X + Y = \int X + \int Y$ can be proved similarly as in the real case. We only need to prove that second part, which involves complex multiplication.

Let $\alpha = a + bi$ and $X(\omega) = U(\omega) + iV(\omega)$, where a and b are real numbers and $U(\omega)$ and $V(\omega)$ are real-valued random variables. Write

$$\alpha X(\omega) = (a + bi)(U(\omega) + iV(\omega))$$
$$= (aU(\omega) - bV(\omega)) + i(aV(\omega) + bU(\omega)).$$

The above equality holds for all $\omega \in \Omega$. Take Lebesgue integral on both sides, we obtain

$$\int \alpha X \triangleq \int (aU - bV) + i \int (aV + bU)$$
$$= a \int U - b \int V + i \left(a \int V + b \int U \right)$$
$$= (a + ib) \left(\int U + i \int V \right) = \alpha \int X.$$

This completes the proof of Theorem 6.7.   ■

We have finished the definition of Lebesgue integral for real-valued and complex-valued and proved the linear and monotonic properties. We will use the Lebesgue integral to define the notion of mathematical expectation of random variable in the next section.

6.4   Mathematical Expectation of Random Variable

When $(\Omega, \mathscr{F}, \mu)$ is a probability space, we define the expectation of X in terms of the Lebesgue integral as follows.

Definition 6.7

The *expectation* of a random variable X is defined by

$$E[X] \triangleq \int X\, dP,$$

where the integral is taken over the sample space Ω with probability measure P.

The use of Lebesgue integration to define the expectation allows for integration over more general measure spaces and a wider range of random variables than the Riemann integral, which is limited to functions on $\mathbb{R}$ or $\mathbb{R}^d$. The expectation of X is also called the *mean* of X. This is an operator that maps a random variable to a real number, satisfying linearity property $E[aX + bY] = aE[X] + bE[Y]$ and monotonic property $X \leq Y \Rightarrow E[X] \leq E[Y]$.

We recover the formulas for the expectation of discrete random variables.

Example 6.4.1 (Expectation of Discrete Random Variable with Finite Range)
Suppose X is a random variable defined on a probability space $(\Omega, \mathcal{F}, P)$ that takes value in $\{a_1, a_2, \ldots, a_n\}$. For $i = 1, 2, \ldots, n$, let $A_i = X^{-1}(\{a_i\})$. We get

$$X(\omega) = \sum_{i=1}^{n} a_i \mathbf{1}_{A_i}(\omega).$$

The expectation of X can be computed as a finite sum

$$E[X] = \int X \, dP = \sum_{i=1}^{n} a_i P(A_i) = \sum_{i=1}^{n} a_i P(X = a_i).$$

When a discrete random variable is integer-valued, we need to consider the positive and negative integers separately.

Example 6.4.2 (Expectation of Integer-Valued Random Variable)
Suppose X is a random variable on the probability space $(\Omega, \mathcal{F}, P)$ whose values are integers. We consider the positive part X^+ and the negative part X^-.

To compute $E[X^+]$, we can take an increasing sequence of simple random variables. For $k = 0, 1, 2, 3, \ldots$, let E_k be the event $\{\omega : 0 \leq X(\omega) \leq k\}$. By the previous example, we obtain

$$\int_{E_k} X^+ \, dP \triangleq \int X^+ \mathbf{1}_{E_k} \, dP = \sum_{i=0}^{k} i P(X = i).$$

Because $X^+ \mathbf{1}_{E_k} \nearrow X^+$ as $k \to \infty$, by the monotone convergence theorem, we get

$$E[X^+] = \lim_{k \to \infty} \sum_{i=0}^{k} i P(X^+ = i) \triangleq \sum_{i=0}^{\infty} i P(X = i).$$

By a similar argument, we obtain

$$E[X^-] = \sum_{\substack{i \in \mathbb{Z} \\ i \leq -1}} (-i) P(X = i) = \sum_{i=1}^{\infty} i P(X = -i).$$

When $E[X^+]$ and $E[X^-]$ are not both equal to infinity, the expectation of X can be computed by

$$E[X] = E[X^+] - E[X^-] = \left(\sum_{i=1}^{\infty} i P(X = i)\right) - \left(\sum_{i=1}^{\infty} i P(X = -i)\right).$$

6.5   Application: Hat Problem and Ball-and-Bin Model

The linearity of expectation

$$E[X_1 + X_2 + X_3 + \cdots + X_n] = E[X_1] + E[X_2] + \cdots + E[X_n]$$

is valid for the sum of integrable random variables, regardless of whether they are independent or not. This gives a simple but effective method in calculating the expectation of random variable in some examples. We decompose the random variable of interest into a sum of indicator functions and use the property that $E[\mathbf{1}_A] = \Pr(A)$.

We give two examples below.

Example 6.5.1 (Hat Problem)
Consider a group of n people who enter a restaurant and put their hats in n boxes. When they leave the restaurant, each person randomly selects a box and takes the hat inside. Let X denote the number of people who retrieve their own hats. A direct calculation of expectation

$$E[X] = \sum_{k} k P(X = k)$$

requires the knowledge of the probability distribution $P(X = k)$. Alternately, We can use the linearity of expectation. For $i = 1, 2, \ldots, n$, let A_i be the event that person i can get back his/her hat back. Writing $X = \sum_{i=1}^{n} \mathbf{1}_{A_i}$, we calculate $E[X]$ by

$$E[X] = \sum_{i=1}^{n} E[\mathbf{1}_{A_i}] = \sum_{i=1}^{n} P(A_i) = n \cdot \frac{1}{n} = 1.$$

Example 6.5.2 (Ball-and-Bin Model)
Consider an experiment in which k indistinguishable balls are thrown independently to n bins. The probability that a ball lands in the i-th bin is $1/n$, for $i = 1, 2, \ldots, n$. Suppose we want to find the expected number of bins that contain exactly one ball. For each $i = 1, 2, \ldots, n$, let A_i be the event that the i-th bin contains exactly one ball. We want to compute the expectation of the sum of indicator functions

$$X = \sum_{i=1}^{n} \mathbf{1}_{A_i}.$$

We note that the events A_i's are dependent, and therefore the indicator functions $\mathbf{1}_{A_i}$, for $i = 1, 2, \ldots, n$, are also dependent. However, the linearity property of expectation holds regardless of dependence between the random variables.
Hence, we have

$$E[X] = \sum_{i=1}^{n} E[\mathbf{1}_{A_i}] = \sum_{i=1}^{n} P(A_i) = \sum_{i=1}^{n} \binom{k}{1} \frac{1}{n} \left(\frac{n-1}{n}\right)^{k-1} = k\left(\frac{n-1}{n}\right)^{k-1}.$$

For fixed n, this expression is maximized when $k = n$. This is because the probability of a ball landing in a bin with exactly one ball is maximized when there is only one ball in each bin, which occurs when $k = n$.

We end this chapter by simulating the ball-and-bin experiment using a Python program. The program generates k independent random numbers, each taking a value in the range $\{0, 1, \ldots, n-1\}$. The sample space of the experiment is thus $\{0, 1, \ldots, n-1\}^k$. To generate the random integers, we use the randint command in Python. A realization is represented by a list of integers called locations.

Next, the program calculates the number of balls in each bin and stores the result in another list of integers called occupancy. This list contains n nonnegative integers. where the i-th number represents the number of balls in bin i. Finally, the program counts the number of entries in occupancy that is equal to 1, which is represented by the variable X.

```
from random import randint

# throw k balls into n bins
n = 10    # number of bins
k = 8     # number of balls

# locations of the k balls
location = [randint(0,n-1) for _ in range(k)]

# compute the number of balls in each bin
occupancy = [sum([x==i for x in locations]) for i in range(n)]

# Find the number of bins with exactly 1 ball
X = sum([y==1 for y in occupancy])

print('location = ', location)
print('occupancy = ', occupancy)
print('Number of bins with exactly one ball: ', X)
```

We note that only the function randint is a source of randomness in this simulation. The other operations are deterministic. The integers in the list occupancy are in fact dependent on each other, for instance, the sum of them should be equal to k. However, without deriving the pmf of the integers in occupancy, we know that the expected value of the random variable X is equal to $k(\frac{n-1}{n})^{k-1}$. For the parameters $k = 8$ and $n = 10$, on average, there are 3.826 bins that contain exactly one ball.

A sample run of this program is as follows:

```
location =   [8, 7, 0, 9, 7, 6, 8, 4]
occupancy =  [1, 0, 0, 0, 1, 0, 1, 2, 2, 1]
Number of bins with exactly one ball:   4
```

In this realization, each of the bins 0, 4, 6, and 9 contains exactly one ball.

Problems

6.1. Consider $\{0, 1, 2, \ldots, n\}$ as a sample space with the discrete σ-algebra. Let p be a real number between 0 and 1. Define a probability measure P by

$$P(\{k\}) = \binom{n}{k} p^k (1 - p)^{n-k},$$

for $k = 0, 1, 2, \ldots, n$. Evaluate the expectation of the following random variables:

(a) $X(k) = k$.

(b) $Y(k) = \begin{cases} 1 & \text{if } k \text{ is even,} \\ -1 & \text{if } k \text{ is odd.} \end{cases}$

6.2. In the hat problem described in Example 6.5.1, find the variance of the number of people who can get back his/her own hat. (Hint: Compute $E[\mathbf{1}_{A_i}\mathbf{1}_{A_j}]$.)

6.3. There are n different types of coupons. In each iteration, a coupon of random type is drawn. Let X_k denote the time you first get k different types of coupons. Trivially, we have $X_1 = 1$. X_2 is the first time that you get a coupon different from the first one. For $k \geq 1$, find the expectation of X_k.

6.4. Let X be a nonnegative random variable that may take ∞ as its value. Show that if $E[X]$ is finite, then X is finite almost surely.

6.5. Let X be an integrable random variable. For any ϵ, show that there is a simple function f such that $E[|X - f|] < \epsilon$.

6.6. Suppose X is a random variable in $L^1(P)$. Show that

$$\lim_{n \to \infty} \int X \mathbf{1}_{|X| \geq n} \, dP = 0.$$

6.7. Consider a sequence $X_1, X_2, X_3, \ldots$ of nonnegative independent random variables, where $P(X_k > a) \geq \delta$ for all k. Here, a is a positive constant and δ is another positive constant between 0 and 1. Show that $\sum_{k=1}^{\infty} X_k = \infty$ with probability 1. (Hint: Use the Borel–Cantelli lemma.)

6.8. Let $(\Omega, \mathscr{F}, \mu)$ be a measure space and X be a nonnegative measurable function defined on this measure space such that $\int_{\Omega} X \, d\mu = 1$. Prove that the set function Q, defined by $Q(A) = \int_A X \, d\mu$ for $A \in \mathscr{F}$, is a probability measure.

6.9. The objective of this question is to prove that for any infinite array of nonnegative real numbers a_{ij}, for $i \geq 1$ and $j \geq 1$, the two different orders of summation give the same answer

$$\sum_{i=1}^{\infty}\sum_{j=1}^{\infty} a_{ij} = \sum_{j=1}^{\infty}\sum_{i=1}^{\infty} a_{ij}. \tag{6.5}$$

(The answer may equal ∞. In this case, both sides are equal to ∞.)

To formulate the problem using the measure theory, we take $\mathbb{N} = \{1, 2, 3, \ldots\}$ as the sample space, equipped with the discrete σ-algebra $\mathscr{P}(\mathbb{N})$. Let μ denote the counting measure on $\mathbb{N}$, i.e., $\mu(A) =$ the size of A for any $A \subseteq \mathbb{N}$.

(a) For each fixed $i \geq 1$, let X_i be the function defined by $X_i(j) = a_{ij}$. Verify that $\int X_i \, d\mu = \sum_{j=1}^{\infty} a_{ij}$.

(b) Let $Y_n = X_1 + X_2 + \cdots + X_n$. Prove that $Y_1, Y_2, Y_3, \ldots$ is an increasing sequence of functions. What is the limit function $Y \triangleq \lim_n Y_n$?

(c) Apply the monotone convergence theorem to prove (6.5).

6.10. Consider the random series

$$S = 1 \pm \frac{1}{2} \pm \frac{1}{4} \pm \frac{1}{8} \pm \frac{1}{16} \pm \frac{1}{32} \pm \cdots,$$

where the signs are chosen independently by tossing a fair coin infinitely many times. Find the mean and the variance of S.

Properties of Lebesgue Integral and Convergence Theorems

7

The Lebesgue integral is known for its superior convergence properties when compared to the Riemann integral. In this chapter, we prove two fundamental convergence theorems related to the Lebesgue integral: the Fatou lemma and the dominated convergence theorem. These properties are not only important in probability theory but also have significant applications in asymptotic analysis for statistics and stochastic processes. However, the Lebesgue integral's more abstract definition makes it impractical to evaluate directly. In the second half of this chapter, we will establish a connection between the Lebesgue integral and the Riemann integral, enabling us to utilize our knowledge of calculus to evaluate a Lebesgue integral more easily.

7.1   Almost-Everywhere Equality

The L^1 norm measures the size of a random variable. We define the L^1 norm of a random variable X as the integral $\int |X| \, d\mu$, and we write the L^1 norm of an integral function X as $\|X\|_1$. The class of integrable function is precisely the class of functions with finite L^1 norm.

The first property is the triangle inequality for integrals, which we will state for both real-valued and complex-valued functions.

Theorem 7.1   (Triangle Inequality for Real Integral and Complex Integral)

$$\left| \int X \, d\mu \right| \leq \int |X| \, d\mu \quad for \ X \in L^1(\mu).$$

© The Author(s), under exclusive license to Springer Nature Switzerland AG 2023
K. Shum, *Measure-Theoretic Probability*, Compact Textbooks in Mathematics,
https://doi.org/10.1007/978-3-031-49830-5_7

Proof When X is $\bar{\mathbb{R}}$-valued, we can prove the inequality easily by using the fact $|X| = X^+ + X^-$,

$$\left| \int X \right| = \left| \int X^+ - \int X^- \right| \leq \int X^+ + \int X^- = \int |X|.$$

A little bit more work is required when X is a complex measurable function. We write X as $U + iV$, where U and V are real-valued functions. If $\int X = 0$, then the inequality is trivially true. Suppose $\int X \neq 0$ and $\int X$ in polar form is $re^{i\theta}$.

Let $\alpha = e^{-i\theta}$. Geometrically, multiplying α and $\int X$ means rotating the point in the complex plane corresponding to $\int X$ by an angle of $-\theta$. Using the properties that the product $\alpha \int X$ is a positive real number and $|\alpha| = 1$, we get

$$\left| \int X \right| = |\alpha| \cdot \left| \int X \right| = \left| \alpha \int X \right| = \left| \int \alpha X \right| = \int \alpha X = \mathrm{Re}\left(\int \alpha X \right) = \int \mathrm{Re}(\alpha X).$$

Since $\mathrm{Re}(z) \leq |z|$ for any complex number z, by monotonic property for real integral, we obtain

$$\left| \int X \right| \leq \int |\alpha X| = \int |\alpha||X| = \int |X|.$$

■

The triangle inequality implies that the L^1 norm satisfies two of the defining properties of a norm function, which are

$$\|X + Y\|_1 \leq \|X\|_1 + \|Y\|_1$$

$$\|\alpha X\|_1 = |\alpha| \cdot \|X\|_1 \quad \text{for any scalar } \alpha.$$

While a function with zero L^1 norm does not necessarily have to be the zero function in general, we can characterize such functions as described in the following theorem.

Theorem 7.2   (Measurable Functions with Zero Norm)
Suppose X is a real-valued or complex-valued measurable function defined on a measure space $(\Omega, \mathscr{F}, \mu)$. Then, the following are equivalent:

1. *$\int_\Omega |X| \, d\mu = 0$.*
2. *X is equal to 0 almost everywhere, i.e., there exists a measurable set A with $\mu(A) = 0$ such that $X(\omega) = 0$ for all $\omega \in A^c$.*

Proof Suppose $X = 0$ almost everywhere. Let A be the set mentioned in the theorem. We have

$$\int |X| = \int_{A^c} |X| + \int_A |X|.$$

The first term is zero because $X(\omega) = 0$ for all $\omega \in A^c$. To show that the second term is zero, we consider nonnegative simple function f that is pointwise smaller than the function $|X|\mathbf{1}_A$. We can write $f(\omega)$ as $\sum_{i=1}^n a_i \mathbf{1}_{E_i}(\omega)$, where E_i is a measurable set inside the set A. Since A has measure 0, all sets E_i's have measure zero, and hence the integral of f is zero. Because it holds for all simple functions $0 \le f \le |X|\mathbf{1}_A$, the Lebesgue integral of $|X|$ over A is zero.

Conversely, suppose $\int |X| = 0$. For each positive integer n, we let E_n denote the event $\{\omega : |X(\omega)| \ge 1/n\}$. The event E_n has measure zero, because

$$0 = \int |X| \, d\mu \ge \int_{E_n} |X| \, d\mu \ge \int_{E_n} (1/n) \, d\mu \ge (1/n)\mu(E_n),$$

which is possible only if $\mu(E_n) = 0$. We can take the set A in the theorem to be the union $\cup_n E_n$, which has measure zero. Then $X(\omega) = 0$ for all $\omega \in A^c$. ■

This theorem says that the integration operator cannot distinguish two functions that differ on a set with measure zero. This motivates us to define an equivalence relation on the set of integrable functions.

Definition 7.1

Consider two integrable functions f and g defined on the same measure space $(\Omega, \mathscr{F}, \mu)$. We say that f and g are *equal everywhere* if $f(\omega) = g(\omega)$ for all $\omega \in \Omega$. We say that f and g are *equal almost everywhere* if there exists an $\mathscr{F}$-measurable set A with $\mu(A) = 0$ such that $f(\omega) = g(\omega)$ for all ω in A^c. We say that two integrable functions are *equivalent* if they are equal almost everywhere. When the measure space is a probability space, we also say that random variables X and Y are equal, *almost surely*, or *with probability 1*, simply a.e., a.s., or w.p.1, if they are equal almost everywhere.

As a concrete example, consider a standard Gaussian random variable $X(\omega)$, and define random variable $Y(\omega)$ to be equal to 10^{100} if $X(\omega)$ is an integer, but is the same as $X(\omega)$ otherwise. If we observe the values of X and Y in practice, they will be equal with probability 1. This is because any difference between X and Y can only occur on a null set, which has probability zero.

The notion of equivalence of random variables depends on the probability measure. It is possible for two random variables to be equivalent under one probability measure but not equivalent under another probability measure. Therefore, when discussing whether two random variables are equal almost surely, we must always specify the underlying probability measure.

In some textbooks, a random variable is defined as a class of equivalent measurable functions, rather than a function itself. In this approach, a measurable function in an equivalence class is called a *version* of the random variable, and the L^1 norm is truly a norm defined on the equivalence classes. That is, if $\int |X| \, d\mu = 0$, then X must belong to the equivalence class of measurable functions that contains the zero function.

However, in this book we will not adopt this interpretation and will instead define a random variable as a measurable function defined on a probability space. Hence, the function $\int |X| \, d\mu$ is only a semi-norm on the space of integrable functions.

This is a matter of convention, and both approaches have their advantages and disadvantages. Defining a random variable as a class of equivalent measurable functions is more convenient mathematically, as we do not need to write the term "a.e." everywhere in the text. However, for the purposes of this book, we will regard a random variable as a measurable function, which is more intuitive and simpler to understand.

Remark 7.1

In view of the fact that Lebesgue integral cannot distinguish two functions that are equal almost everywhere, we can further extend the notion of Lebesgue integral. Given a function f, which may or may not be measurable, we say that it is *integrable* if it is equal a.e. to a function g that is measurable and integrable, and we define the integral of f as the same as the Lebesgue integral of g. This extension is useful to compute the integral for function that is integrable outside a set of measure zero, and is not well defined inside.

7.2   Fatou's Lemma and Dominated Convergence Theorem

From now on we will only require that the conditions about a random variable in a theorem are satisfied almost everywhere. This will make the theorems more general. In fact, the monotone convergence theorem (Theorem 6.4) remains to be true if the conditions in this theorem are satisfied almost everywhere.

The Fatou lemma below only assumes that the sequence of measurable functions is nonnegative, without assuming that the sequence is monotonically increasing, and hence the assumption is much more relaxed than that in the monotone convergence theorem. However, the conclusion is also weaker.

Theorem 7.3 (Fatou's Lemma)
Let X_n, $n = 1, 2, 3, \ldots$, be a sequence of measurable functions defined on a measure space $(\Omega, \mathscr{F}, \mu)$, such that each of them is nonnegative almost everywhere. Then

$$\int \liminf_{n \to \infty} X_n \, d\mu \leq \liminf_{n \to \infty} \int X_n \, d\mu.$$

Proof The function $\liminf_{n \to \infty} X_n(\omega)$ is well defined and is nonnegative almost everywhere. To be precise, we can assume that each X_n is nonnegative on a measurable set E_n with $\mu(E_n^c) = 0$, for $n = 1, 2, 3, \ldots$. Then, we can consider the intersection $E = \cap_{n=1}^{\infty} E_n$, which is measurable and satisfies $X_n(\omega) \geq 0$ for all n and all ω in this set. Moreover, the complement of the set E has measure zero.

For $n \geq 1$, define $Y_n(\omega) \triangleq \inf_{k \geq n} X_k(\omega)$. The sequence $Y_1 \leq Y_2 \leq Y_3 \leq \cdots$ is non-decreasing and is converging pointwise on the set E. By the monotone convergence theorem (Theorem 6.4),

$$\int_E Y_n \, d\mu \nearrow \int_E \lim_{n \to \infty} Y_n \, d\mu = \int_E \liminf_n X_n \, d\mu.$$

Therefore

$$\sup_n \int_E Y_n \, d\mu = \int_E \liminf_n X_n \, d\mu. \tag{7.1}$$

We then use the definition of $Y_n = \inf_{k \geq n} X_k$ to see that $X_k \geq Y_n$ on E for all $k \geq n$. Hence, by the monotonic property of integral, we get

$$\int_E X_k \, d\mu \geq \int_E Y_n \, d\mu \quad \text{for all } k \geq n.$$

Combining with the fact that $\mu(E^c) = 0$, we obtain

$$\sup_n \left(\inf_{k \geq n} \int_\Omega X_k \, d\mu \right) \geq \int_\Omega \liminf_n X_n \, d\mu.$$

This proves the inequality in Fatou's lemma. ∎

Example 7.2.1 (An Example That Illustrates Fatou's Lemma)
We give an example to help us remember the direction of the inequality in Fatou's lemma. Consider an example with sample space $\Omega = \mathbb{R}$ with the Lebesgue measure on $\mathbb{R}$. For $n \geq 1$, let X_n be the simple function $n\mathbf{1}_{[0,1/n]}$. The integral of X_n over $\mathbb{R}$ is equal to 1 for all n. Therefore $\liminf_n \int X_n \, d\mu = 1$.

However for each nonzero $\omega \in \mathbb{R}$, $\liminf_n X_n(\omega) = 0$, and at the point $\omega = 0$, the liminf is equal to ∞. Thus, we see that $X(\omega) = \limsup_n X_n(\omega)$ is equal to

$$
X(\omega) = \begin{cases} 0 & \text{if } \omega \neq 0 \\ \infty & \text{if } \omega = 0. \end{cases}
$$

This yields $\int X \, d\mu = 0$. Informally, Fatou's lemma says that the integral becomes larger if we pull the liminf operator outside of the integral sign.

The next example demonstrates the importance of understanding the conditions under which the limit and integral can be interchanged.

Example 7.2.2 (Limit of Riemann Integrable Functions May Not Be Integrable)
We can construct a sequence of functions on $\mathbb{R}$ by enumerating the rational numbers between $[0, 1]$. Since the rational numbers are countable, we can list them as $r_1, r_2, r_3, \ldots$. For $i = 1, 2, 3, \ldots$, we let $f_i(x) = \sum_{k=1}^{i} \mathbf{1}_{\{r_k\}}(x)$ be the sum of indicator functions. Each f_i is Riemann integrable on $[0, 1]$, and its integral over this interval is equal to 0.

However, the sequence of functions $(f_i)_{i=1}^{\infty}$ converges pointwise to the indicator function $\mathbf{1}_{\mathbb{Q} \cap [0,1]}(x)$, which is not Riemann integrable on $[0, 1]$. This means that we have a sequence of Riemann integrable functions whose integrals converge to 0, but the limit of the functions is not even Riemann integrable.

Theorem 7.4 (Dominated Convergence Theorem)
For $n = 1, 2, 3, \ldots$, let $X_n(\omega)$ be $\bar{\mathbb{R}}$-valued measurable functions defined on a measure space $(\Omega, \mathscr{F}, \mu)$, converging almost everywhere to $X(\omega)$,

$$
X_n \to X \quad a.e.
$$

If there exists a nonnegative integrable function Y such that $|X_n| \leq Y$ a.e. for all n, then

(1) X is Lebesgue integrable, i.e., $X \in L^1(\mu)$, and
(2)

$$
\lim_{n \to \infty} \int X_n \, d\mu = \int X \, d\mu = \int \lim_{n \to \infty} X_n \, d\mu.
$$

Proof By the assumption that $|X_n| \leq Y$ a.e., we can find for each n a measurable set E_n such that $\mu(E_n^c) = 0$ and $|X_n(\omega)| \leq Y(\omega)$ for all $\omega \in E_n$. Also, we can find a measurable set E_0 such that $\mu(E_0^c) = 0$ and $X_n(\omega) \to X(\omega)$ for all $\omega \in E_0$. For ω in the set $E \triangleq \cap_{n \geq 0} E_n$, we have $|X_n(\omega)| \leq Y(\omega)$ and $X_n(\omega) \to X(\omega)$.

We next observe that X_n is integrable for all n, since $\int |X_n| \leq \int Y < \infty$.

Although we do not know the value of $X(\omega)$ for $\omega \in E_0^c$, we may set $X(\omega)$ to 0 for $\omega \in E_0^c$ without loss of generality. Since $X_n(\omega)$ converges to $X(\omega)$ for all $\omega \in E_0$ by assumption, the function X is a measurable function. (See the remark at the end of Section 7.1)

Consider the difference $|X_n - X|$. By applying the triangle inequality for real numbers, we have

$$|X_n(\omega) - X(\omega)| \leq |X_n(\omega)| + |X(\omega)| \leq 2Y(\omega)$$

for each $\omega \in E$. Hence $2Y - |X_n - X| \geq 0$ almost everywhere. By Fatou's lemma,

$$\int \liminf_n (2Y - |X_n - X|) \leq \liminf_n \int 2Y - |X_n - X|. \tag{7.2}$$

Since $|X_n - X| \to 0$ as $n \to \infty$, the left-hand side equals $\int 2Y$, and the right-hand side becomes

$$\liminf_n \left(\int 2Y + \int -|X_n - X| \right) = \int 2Y - \limsup_n \int |X_n - X|.$$

In the last step, we have pulled out the constant $\int 2Y$, which does not depend on n, and change liminf to limsup. Because $\int 2Y$ is finite, we can subtract it from both sides of (7.2) to obtain

$$0 \geq \limsup_n \int |X_n - X|.$$

The limsup of a sequence of nonnegative real numbers $\int |X_n - X|$ should be nonnegative. Hence, $\limsup_n \int |X_n - X| = 0$, and this is possible only when $\int |X_n - X|$ is actually converging to 0,

$$\lim_{n \to \infty} \int |X_n - X| = 0. \tag{7.3}$$

By the triangle inequality for real numbers and monotonic property for integrals,

$$\int |X| = \int |X - X_n + X_n| \leq \int |X - X_n| + \int |X_n|,$$

which holds for any positive integer n. By (7.3), for any arbitrarily small $\epsilon > 0$, we can choose a sufficiently large n such that $\int |X - X_n| < \epsilon$. We then use the assumption that $|X_n| \leq Y$ to see that $\int |X| \leq \epsilon + \int |Y| < \infty$. Therefore, by Theorem 6.6, X is integrable. This proves (1).

We apply the triangle inequality (Theorem 7.1) to get

$$\left| \int X_n - X \, d\mu \right| \leq \int |X_n - X| \, d\mu.$$

Because $\int |X_n - X| \, d\mu \to 0$ as $n \to \infty$, we have $|\int X_n - X \, d\mu| \to 0$ as $n \to \infty$. This is the same as saying that $\int X_n \, d\mu \to \int X \, d\mu$ as $n \to \infty$. This proves the equalities in (2) and completes the proof of the dominated convergence theorem. ∎

We highlight the intermediate step in (7.3) that is important by itself.

Theorem 7.5 (L^1 Convergence)
With the conditions in Theorem 7.4, we have

$$\int |X_n - X| \, d\mu \to 0, \qquad as \ n \to \infty.$$

In fact, we have shown that the convergence of the L^1 norm of $X_n - X$ to zero is sufficient to derive the conclusions of the dominated convergence theorem.

Example 7.2.3 (A Counterexample to the Dominated Convergence Theorem)
The assumption that $|X_n| \leq Y$ in the dominated convergence theorem cannot be relaxed. For $n = 1, 2, 3, \ldots$, consider the indicator function $\mathbf{1}_{[n,n+1]}(x)$ defined on the real number line. We put the Borel algebra and the Lebesgue measure μ on the sample space $\mathbb{R}$. There is no integrable function Y that can dominate all the X_n's. The conclusion in the dominated convergence theorem fails because, although the functions X_n converge pointwise to the zero function, we have

$$\lim_{n \to \infty} \int_{\mathbb{R}} X_n \, d\mu = 1 \neq 0 = \int_{\mathbb{R}} \lim_{n \to \infty} X_n \, d\mu.$$

The dominated convergence theorem can be extended readily to the complex case.

Theorem 7.6 (Complex DCT, Sequential Version)
Let X_n be complex-valued random variables, for $n = 1, 2, 3, \ldots$, defined on a measure space $(\Omega, \mathcal{F}, \mu)$. Suppose $(X_n)_{n \geq 1}$ converges to X almost everywhere for all n, and there exists an integrable function Y such that $|X_n| \leq Y$ almost everywhere for all n. Then $X \in L^1(\mu)$ and

$$\lim_{n \to \infty} \int X_n \, d\mu = \int \lim_{n \to \infty} X_n \, d\mu.$$

The proof follows the same steps as in the proof of Theorem 7.4. We just need to apply the real version of DCT to the real parts and imaginary parts.

Another useful form of DCT is for random variables indexed by a continuous variable. One takes limit as the index variable approaches certain limit.

Theorem 7.7 (Complex DCT, Limit Version)
Let X_h be complex-valued measurable functions defined on a measure space $(\Omega, \mathcal{F}, \mu)$, indexed by a real variable h in some interval. Suppose $\lim_{h \to h_0} X_h(\omega)$ converges almost everywhere for some h_0, and assume that there exists a real-valued integrable function Y such that $|X_h| \leq Y$ for all h in the interval. Then the limit function $X(\omega) \triangleq \lim_{h \to h_0} X_h(\omega)$ is integrable and

$$\lim_{h \to h_0} \int X_h \, d\mu = \int \lim_{h \to h_0} X_h \, d\mu.$$

The proof of Theorem 7.7 can be reduced to the sequential version of DCT by constructing a sequence of random variables converging to X. The full proof is omitted here.

7.3 Application: Evaluation of Lebesgue–Stieltjes Integrals

As an application of the dominated convergence theorem, we provide a sufficient condition under which a Lebesgue integral on the real number line can be computed as a Riemann–Stieltjes integral. This connection allows us to evaluate abstract Lebesgue–Stieltjes (LS) integral using tools from calculus. Throughout the rest of this section, we denote the Riemann–Stieltjes (RS) integral of a function $g(x)$ with respect to a Stieltjes measure function $F(x)$ from a to b by $\int_a^b g(x) \, dF(x)$.

The next theorem is a link between the Lebesgue–Stieltjes integral and the Riemann–Stieltjes integral.

Theorem 7.8
Consider a measure space $(\mathbb{R}, \mathscr{B}(\mathbb{R}), \mu)$ associated with the Stieltjes measure function $F(x) = \mu((-\infty, x])$. Assume that $\mu(\{a\}) = 0$, i.e., assume $F(x)$ is continuous at $x = a$. If g is a continuous function on $[a, b]$, then g is μ-integrable on $[a, b]$ and we have

(continued)

Theorem 7.8 (continued)

$$\int_{[a,b]} g \, d\mu = \int_a^b g(x) \, dF(x),$$

where the left-hand side is an LS integral and the right-hand side is an RS integral.

Proof Since g is continuous on $[a, b]$, it is measurable. Furthermore, g is bounded on $[a, b]$, i.e., there exists a constant K such that $|g(x)| \leq K$ for all $x \in [a, b]$. Therefore, the function g is μ-integrable because

$$\int_{[a,b]} g^+ \, d\mu \leq \int_{[a,b]} K \, d\mu = K\mu([a, b]) < \infty$$

$$\int_{[a,b]} g^- \, d\mu \leq \int_{[a,b]} K \, d\mu = K\mu([a, b]) < \infty.$$

By Theorem 1.1, the Riemann–Stieltjes on the right-hand side exists since g is continuous.

The idea of proof is to approximate the function g by simple functions. We partition $[a, b]$ into mutually disjoint intervals

$$\{a\}, \ (a_1, b_1], \ (a_2, b_2], \dots, (a_n, b_n],$$

with $a = a_1 < b_1 = a_2 < b_2 = \cdots a_n < b_n = b$, and define two simple functions

$$\underline{g}(x) \triangleq g(a)\mathbf{1}_{\{a\}} + \sum_{i=1}^n m_i \mathbf{1}_{(a_i, b_i]}, \qquad \overline{g}(x) \triangleq g(a)\mathbf{1}_{\{a\}} + \sum_{i=1}^n M_i \mathbf{1}_{(a_i, b_i]},$$

where

$$m_i \triangleq \inf_{a_i < x \leq b_i} g(x) \text{ and } M_i \triangleq \sup_{a_i < x \leq b_i} g(x)$$

for $i = 1, 2, \dots, n$. It is easily seen that

$$\underline{g} \leq g \leq \overline{g}.$$

Compute the Lebesgue integrals of $\underline{g}$, g, and $\overline{g}$ and apply the monotonic property,

$$\sum_{i=1}^n m_i \mu((a_i, b_i]) = \int_{[a,b]} \underline{g} \leq \int_{[a,b]} g \leq \int_{[a,b]} \overline{g} \leq \sum_{i=1}^n M_i \mu((a_i, b_i]).$$

We have used the assumption that $\mu(\{a\}) = 0$ in this step.

Because the Riemann–Stieltjes integral $\int_a^b g(x)\,dF(x)$ exists, the left-most and right-most sums above converge to a common limit. By a sandwich argument, the integral $\int_{[a,b]} g\,d\mu$ in the middle is equal to the Riemann–Stieltjes integral $\int_a^b g(x)\,dF(x)$.   ∎

▶ **Practical Calculations of Lebesgue-Stieltjes Integrals**   Theorem 7.8 assumes that g is continuous, but this assumption can be relaxed. With a little more work, one can show that:

(a) A function g is Riemann–Stieltjes integrable if and only if g is continuous almost everywhere with respect to the measure μ on $[a, b]$.
(b) If g is Riemann–Stieltjes integrable on $[a, b]$, then g is integrable with respect to the completion of the measure μ induced by the Stieltjes measure function F, and the two integrals are equal to each other.

Nonetheless, Theorem 7.8 is sufficient for many practical calculations. If the function g to be integrated is piece-wise continuous, we can compute the integral by dividing the x-axis into multiple pieces so that g is continuous in each part. As long as the discontinuity points of $g(x)$ and the discontinuity points of the Stieltjes measure function $F(x)$ do not overlap, we can compute the integral of $g(x)$.

We note that the result in Theorem 7.8 also covers Lebesgue integration with respect to the Lebesgue measure λ. The Lebesgue integral $\int_{[a,b]} g\,d\lambda$ can be evaluated using the Riemann integral $\int_a^b g(x)\,dx$.

Example 7.3.1 (Computation of Lebesgue–Stieltjes Integral)
Consider a probability space $(\mathbb{R}, \mathscr{B}(\mathbb{R}), P)$, where P is the probability measure generated by the Stieltjes measure function

$$F(x) = \begin{cases} 0 & \text{if } x \le 0 \\ x^2 & \text{if } 0 < x \le 1 \\ 1 & \text{if } 1 \le x. \end{cases}$$

The Lebesgue integral of a continuous function $g(x)$ defined on interval $[0, 1]$ can be evaluated by

$$\int_{[0,1]} g(x)\,dP(x) = \int_0^1 g(x)\,d(x^2).$$

When $g(x) = x$, it is the expectation of the probability distribution

$$\int_0^1 x\,d(x^2) = \int_0^1 2x^2\,dx = \frac{2}{3}[x^3]_0^1 = \frac{2}{3}.$$

If we want to compute $\int \sin(\pi x)\,dP(x)$, we can first reduce it to a Riemann integral

$$\int \sin(\pi x)\,dP(x) = \int_0^1 \sin(\pi x)\,d(x^2) = 2\int_0^1 x\sin(\pi x)\,dx.$$

Using integration by parts, we obtain the answer $2/\pi$.

Theorem 7.9

We continue with the notation in Theorem 7.8. Suppose g is a piece-wise continuous function so that the Riemann–Stieltjes integral

$$\int_a^b g(x) \, dF(x) \quad \text{exists for all } a < b.$$

If one of the integrals

$$\lim_{\substack{b \to \infty \\ a \to -\infty}} \int_a^b |g(x)| \, dF(x) \text{ or } \int_{\mathbb{R}} |g(x)| \, d\mu(x) \tag{7.4}$$

exists and is finite, then the other integral exists and is finite. In this case, we have

$$\int_{\mathbb{R}} g(x) \, d\mu(x) = \int_{-\infty}^{\infty} g(x) \, dF(x). \tag{7.5}$$

Proof We first suppose that $\int_{-\infty}^{\infty} g(x) \, dF(x)$ exists and prove that $g(x)$ is μ-integrable. We truncate the function $g(x)$ to $g(x) \cdot \mathbf{1}_{[-n,n]}(x)$, whose value is zero when $|x| > n$. By the monotone convergence theorem,

$$\int_{\mathbb{R}} |g(x)| \, d\mu = \lim_{n \to \infty} \int |g(x)| \mathbf{1}_{[-n,n]} \, d\mu = \lim_{n \to \infty} \int_{-n}^{n} |g(x)| \, dF(x) < \infty.$$

We have applied Theorem 7.8 in the second step above. Therefore $g \in L^1(\mu)$.

Conversely, suppose that $g(x)$ is μ-integrable. This is equivalent to $\int_{\mathbb{R}} |g| \, d\mu < \infty$. Therefore, for any finite $a < b$,

$$\int_a^b |g(x)| \, dF(x) = \int_{[a,b]} |g(x)| \, d\mu(x) \leq \int_{\mathbb{R}} |g(x)| \, d\mu(x).$$

Since g is in $L^1(\mu)$, the Lebesgue integral $\int_{\mathbb{R}} |g|$ is finite. So $\int_a^b |g(x)| \, dF(x)$ is upper bounded by a constant, for all a and for all b. Taking limit, the integral $\int_{-\infty}^{\infty} |g(x)| \, dF(x)$ is also upper bounded by the same constant and hence is finite.

Next, assume that one (and hence both) of the integrals in (7.4) is finite. Apply DCT with $X_n \triangleq g \cdot \mathbf{1}_{[-n,n]}$ and $Y = |g|$. We check that $|X_n| \leq Y$ and $X_n \to g$ as $n \to \infty$. Since Y is integrable by assumption, we can apply DCT to obtain

$$\int_{\mathbb{R}} g(x) \, d\mu(x) = \lim_{n \to \infty} \int_{\mathbb{R}} g(x) \cdot \mathbf{1}_{[-n,n]}(x) \, d\mu(x)$$

$$= \lim_{n \to \infty} \int_{-n}^{n} g(x) \, dF(x) \triangleq \int_{-\infty}^{\infty} g(x) \, dF(x).$$

This proves that the LS integral and the improper RS integral in (7.5) have the same value.   ∎

We will demonstrate the application of the previous theorem by using it to calculate the mean of the exponential distribution, which has infinite support.

Example 7.3.2 (Expected Value of the Exponential Distribution)
The probability density function of the exponential distribution is given by

$$f(x) \triangleq \lambda e^{-\lambda x} \cdot \mathbf{1}_{[0,\infty)}(x),$$

where λ is a positive constant. We can create an LS measure on $\mathbb{R}$ from the Stieltjes measure function

$$F(x) = \begin{cases} \int_0^x \lambda e^{-\lambda t} \, dt & \text{if } x \geq 0, \\ 0 & \text{if } x < 0. \end{cases}$$

Denote the resulting LS measure by P. Suppose we want to compute the mean $\int_{\mathbb{R}} x \, dP(x)$. Since the negative real number line has measure zero with respect to measure P, we may focus on the positive real number line. To apply Theorem 7.9, we check that the RS integral

$$\int_a^b x \, dF(x) = \int_a^b x \lambda e^{-\lambda x} \, dx$$

exists for all $0 \leq a < b$, and moreover, the improper integral $\int_0^\infty x \lambda e^{-\lambda x} \, dx$ exists and is equal to $1/\lambda$, which is a finite value. Therefore $\int_{[0,\infty)} x \, dP(x) = 1/\lambda$.

7.4   Push-Forward Measure and Change-of-Variable Formula

When modeling a stochastic experiment, we often assume that a sample is drawn from a probability space according to a probability measure. However, the probability space may be complex, and evaluating a Lebesgue integral on it may not always be straightforward. In some cases, we cannot directly observe the sample drawn from the sample space, but instead we measure the experiment using random variables.

To calculate the expectation of these random variables, it is often more convenient to transform the original probability space to a more concrete probability space defined on the real number line. If the Stieltjes measure function is differentiable, we can simplify a Lebesgue integral on the real line by transforming it to a Riemann integral. This simplification is useful, as Riemann integrals are often easier to compute than Lebesgue integrals. The reduction process is shown as follows:

Lebesgue integral on→	Lebesgue–Stieltjes→	Riemann–Stieltjes→	Riemann
abstract prob. space	integral on $\mathbb{R}$	integral on $\mathbb{R}$	integral

In this section we consider an abstract measure space $(\Omega, \mathscr{F}, \mu)$ and a real-valued measurable function X that maps out of it.

$$(\Omega, \mathscr{F}, \mu) \xrightarrow{X} (\mathbb{R}, \mathscr{B}(\mathbb{R})). \tag{7.6}$$

Theorem 7.10
Given a measurable mapping X as in (7.6), the set function

$$X_{\#}\mu(B) \triangleq \mu(X^{-1}(B)) = \mu(\{\omega : X(\omega) \in B\})$$

is a Borel measure defined for all $B \in \mathscr{B}(\mathbb{R})$.

Proof Since X is $(\mathscr{F}, \mathscr{B}(\mathbb{R}))$-measurable, the pre-image $X^{-1}(B)$ is $\mathscr{F}$-measurable. Hence $X_{\#}\mu$ is well-defined. We check the axioms of measure below. It is obvious that $X_{\#}\mu(\emptyset) = 0$. For any sequence of mutually disjoint sets $B_1, B_2, \ldots$ in $\mathscr{B}(\mathbb{R})$,

$$X_{\#}\mu(\uplus_{i=1}^{\infty} B_i) = \mu(X^{-1}(\uplus_{i=1}^{\infty} B_i)) = \mu(\uplus_{i=1}^{\infty} X^{-1}(B_i)) = \sum_{i=1}^{\infty} X_{\#}\mu(B_i).$$

The second equality follows from (4.3) and the third equality from the assumption that μ is a measure function. ∎

Definition 7.2

The measure $X_{\#}\mu$ defined in Theorem 7.10 is called the *push-forward* measure defined by X. It is also called the *image* of μ or the *measure induced* by X. When μ is a probability measure, the induced measure $X_{\#}\mu$ is also a probability measure and is called the *distribution* of X.

Other notation for the push-forward measure include $X_{*}\mu$, μ^{X}, μ_{X}, and $\mu \circ X^{-1}$.

Example 7.4.1 (Rayleigh Distribution)
We generate a random point on $\mathbb{R}^2$, so that the x- and y-coordinates are independent and identically distributed Gaussian random variables. Suppose the mean is 0 and the variance is σ^2. We want to find the distribution of the distance of the random point to the origin.

We can formulate this problem by regarding $(\mathbb{R}^2, \mathscr{B}(\mathbb{R}), P)$ as the probability space and specify the probability measure P on the semi-infinite rectangles

$$P((-\infty, x] \times (-\infty, y]) = \int_{-\infty}^{x} \frac{e^{-x^2/(2\sigma^2)}}{\sqrt{2\pi\sigma^2}} \, dx \int_{-\infty}^{y} \frac{e^{-y^2/(2\sigma^2)}}{\sqrt{2\pi\sigma^2}} \, dy.$$

The measure P will then be uniquely defined by the measure extension theorem, as the Borel algebra on $\mathbb{R}^2$ is generated by the semi-infinite rectangles.

Let Z denote the distance between the random point and the origin. The distance Z is less than z if and only if $X^2 + Y^2 \leq z^2$. Hence, we can derive the cdf of Z by

$$P(Z \leq z) = P(X^2 + Y^2 \leq z^2) = \frac{1}{2\pi\sigma^2} \iint_{x^2+y^2 \leq z^2} e^{-(x^2+y^2)/(2\sigma^2)} \, dx dy.$$

After some calculus, we can simplify it to $P(Z \leq z) = 1 - e^{-z^2/(2\sigma^2)}$. The pushforward measure $Z_\# P$ is the probability measure on $\mathbb{R}$ induced by the Stieltjes measure function $F(z) = 1 - e^{-z^2/(2\sigma^2)}$, which is the cdf of a Rayleigh distribution.

The change-of-variable formula gives the relation between a Lebesgue integral on an abstract space and a Lebesgue integral on the real number line.

Theorem 7.11 (Change-of-Variable Formula)
Let $(\Omega, \mathscr{F}, \mu)$ be a measure space, and consider two measurable maps X and h

$$(\Omega, \mathscr{F}, \mu) \xrightarrow{X} (\mathbb{R}, \mathscr{B}(\mathbb{R})) \xrightarrow{h} (\mathbb{R}, \mathscr{B}(\mathbb{R})).$$

Let $X_\#\mu$ denote the measure on $\mathbb{R}$ induced by X. Then,

(a) $h(X) \in L^1(\mu)$ iff $h \in L^1(X_\#\mu)$.
(b) If h is nonnegative or if the two equivalent conditions in (a) hold, then

$$\int_\Omega h(X(\omega)) \, d\mu(\omega) = \int_\mathbb{R} h(x) \, dX_\#\mu(x). \qquad (7.7)$$

The function h in the theorem may be continuous or piece-wise continuous. The important requirement is that the composite function $h(X)$ is measurable, so that the composite function $h \circ X$ is also measurable.

Proof Suppose h is an indicator function $\mathbf{1}_B$ for some Borel set $B \in \mathscr{B}(\mathbb{R})$. The left-hand side of (7.7) is

$$\int_\Omega h(X(\omega)) \, d\mu(\omega) = \int_\Omega (\mathbf{1}_B \circ X)(\omega) \, d\mu(\omega) = \int_\Omega \mathbf{1}_{X^{-1}(B)} \, d\mu = \mu(X^{-1}(B)).$$

Meanwhile, the right-hand side is

$$\int_\mathbb{R} \mathbf{1}_B \, dX_\#\mu = X_\#\mu(B) = \mu(X^{-1}(B)).$$

The first equality above follows from the definition of Lebesgue integral for simple function and the second from the definition of $X_\# \mu$. Therefore (7.7) holds for indicator functions h.

Suppose h is a nonnegative and measurable function. Let h_n be simple and nonnegative functions for $n = 1, 2, 3, \ldots$, such that $h_n \nearrow h$. We have $h_n(X) \nearrow h(X)$. Apply the monotone convergence theorem two times in the following derivation:

$$
\int_\Omega h(X) \, d\mu = \int_\Omega \lim_{n \to \infty} h_n(X(\omega)) \, d\mu(\omega)
$$

$$
\stackrel{\text{MCT}}{=} \lim_{n \to \infty} \int_\Omega h_n(X(\omega)) \, d\mu(\omega)
$$

$$
= \lim_{n \to \infty} \int_{\mathbb{R}} h_n(x) \, dX_\# \mu(x)
$$

$$
\stackrel{\text{MCT}}{=} \int_{\mathbb{R}} \lim_{n \to \infty} h_n(x) \, dX_\# \mu(x)
$$

$$
= \int_{\mathbb{R}} h(x) \, dX_\# \mu(x).
$$

Therefore (7.7) holds for nonnegative and measurable functions.

Next we assume that h is a real-valued and measurable function. Applying the argument in the previous paragraph to $|h|$, we obtain part (a) immediately.

Write $h = h^+ - h^-$, where h^+ and h^- are the positive and negative parts of h. Suppose $h(X)$ is integrable, i.e., suppose that $\int h^+(X) < \infty$ and $\int h^-(X) < \infty$. We note that $h^+(X)$ and $h^-(X)$ are both nonnegative and measurable functions in $L^1(\mu)$. The integral of $h(X)$ is

$$
\int_\Omega h(X) \, d\mu \triangleq \int_\Omega h^+(X(\omega)) \, d\mu(\omega) - \int_\Omega h^-(X(\omega)) \, d\mu(\omega)
$$

$$
= \int_{\mathbb{R}} h^+(x) \, dX_\# \mu(x) - \int_{\mathbb{R}} h^-(x) \, dX_\# \mu(x)
$$

$$
= \int_{\mathbb{R}} h \, dX_\# \mu.
$$

This finishes the proof of the change-of-variable formula.   ∎

When applied to finite probability spaces, the change-of-variable formula says that we can compute a function of a random variable X by

$$
E[g(X)] = \sum_{i=0}^{\infty} g(i) p_i,
$$

where p_i is the probability $P(X = i)$, for $i = 0, 1, 2, \ldots$, and g is a real-valued function.

In the next theorem, the formula given in Eq. (7.8) has earned the nickname "LOTUS" (Law of the Unconscious Statistician). Additionally, the formula in Eq. (7.9) serves as the definition of the expectation of a continuous random variable in a first course in probability. We thus see that the measure-theoretic definition of expectation unifies the notion of mathematical expectation for discrete and continuous random variables.

Theorem 7.12

Let X be a random variable defined on $(\Omega, \mathscr{F}, P)$. Suppose the induced measure $X_\# P$ on $\mathbb{R}$ has Stieltjes measure function $F_X(x)$, and suppose $F_X(x)$ is differentiable with derivative $f_X(x)$. Then, for any piece-wise continuous function $g(x)$ such that $g(X)$ is P-integrable, $E[g(X)]$ is given by

$$E[g(X)] = \int_{-\infty}^{\infty} g(x) f_X(x)\, dx. \tag{7.8}$$

In particular, when the function $g(x) = x$ is the identity function, we recover the formula for computing the mean of a continuous random variable

$$E[X] = \int_{-\infty}^{\infty} x f_X(x)\, dx. \tag{7.9}$$

Proof By the change-of-variable formula in (7.7), the expectation of $g(X)$ is equal to the Lebesgue–Stieltjes integral

$$\int_{\Omega} g(X)\, dP = \int_{\mathbb{R}} g(x)\, dX_\# P(x).$$

We can transform it to a Riemann–Stieltjes integral

$$\int_{-\infty}^{\infty} g(x)\, dF_X(x)$$

by applying Theorem 7.9. Since $F_X(x)$ is differentiable, this can be further simplified to a Riemann integral $\int_{-\infty}^{\infty} g(x) f_X(x)\, dx$. ∎

One immediate application of LOTUS is the computation of the mean and variance of a probability density distribution.

Example 7.4.1 (Variance of a Random Variable)
Let X be a discrete random variable with pmf $P(X = i) = p_i$ for $i \geq 0$ and mean m. To evaluate the variance of X, we need not obtain the distribution of random variable $Y = (X - m)^2$, but rather compute it directly using the formula

$$E[(X - m)^2] = \sum_i (i - m)^2 p_i.$$

Example 7.4.2 (Computing the Mean of a Chi Distribution)
The chi distribution with k degree of freedom is the square root of the chi-square distribution with the same degree of freedom. Although the pdf of the chi distribution is available in closed form, in the computation of the mean, however, we can simply use the pdf of chi-square distribution

$$f_{\chi^2}(x) = \begin{cases} \frac{1}{2^{k/2}\Gamma(k/2)} x^{k/2-1} e^{-x/2} & \text{if } x \geq 0 \\ 0 & \text{if } x < 0, \end{cases}$$

which is a special case of Gamma distribution. By Theorem 7.12, we can write

$$\int_0^\infty \sqrt{x} f_{\chi^2}(x)\, dx = \int_0^\infty \sqrt{x} \frac{1}{2^{k/2}\Gamma(k/2)} x^{k/2-1} e^{-x/2}\, dx$$

$$= \frac{\sqrt{2}\Gamma((k+1)/2)}{\Gamma(k/2)} \int_0^\infty \frac{1}{2^{(k+1)/2}\Gamma((k+1)/2)} x^{\frac{k+1}{2}-1} e^{-x/2}\, dx.$$

The integral is equal to 1 because the summand is the pdf of a Gamma distribution. The expectation of the chi distribution with k degree of freedom equals $\frac{\sqrt{2}\Gamma((k+1)/2)}{\Gamma(k/2)}$.

7.5   Expectation of the Product of Two Independent Random Variables

Theorem 7.13
Suppose X and Y are random variables defined on a common probability space $(\Omega, \mathcal{F}, P)$. If X and Y are independent and if $E[X]$, $E[Y]$, and $E[XY]$ are all finite, then

$$E[XY] = E[X]E[Y]. \tag{7.10}$$

Proof We first suppose that X and Y are indicator functions. Suppose $X = \mathbf{1}_A$ and $Y = \mathbf{1}_B$ for some sets A and B. Since X and Y are measurable, the sets A and B are both $\mathcal{F}$-measurable. Moreover, since X and Y are independent, the events A and B are independent. Therefore,

$$E[\mathbf{1}_A \mathbf{1}_B] = E[\mathbf{1}_{A \cap B}] = P(A \cap B) = P(A)P(B) = E[\mathbf{1}_A]E[\mathbf{1}_B].$$

By linearity, (7.10) holds when X and Y are independent simple functions.

Suppose X and Y are nonnegative, measurable, and independent random variables. Let X_n's and Y_n's be simple nonnegative functions obtained by the method mentioned after Definition 6.3. For each n, X_n and Y_n are independent because X_n is a function of X and Y_n is a function of Y, and X and Y are independent (Theorem 5.2). Furthermore, we have $X_n Y_n \nearrow XY$ because $X_n \nearrow X$ and $Y_n \nearrow Y$. By the monotone convergence theorem,

$$E[XY] = \lim_n E[X_n Y_n] = \lim_n (E[X_n]E[Y_n]) = (\lim_n E[X_n])(\lim_n E[Y_n]),$$

which is the same as $E[X]E[Y]$.

Suppose X and Y are in $L^1(P)$. Let X^+ and X^- denote the positive and negative parts of X, respectively, and let Y^+ and Y^- denote the positive and negative parts of Y, respectively. We note that the pair of random variables (X^+, X^-) is independent with (Y^+, Y^-) because $X^+ = \max(X, 0)$ and $X^- = \max(-X, 0)$ are functions of X and $Y^+ = \max(Y, 0)$ and $Y^- = \max(-Y, 0)$ are functions of Y (Theorem 5.2). Then, we can write

$$E[XY] = E[(X^+ - X^-)(Y^+ - Y^-)]$$
$$= E[X^+]E[Y^+] - E[X^+]E[Y^-] - E[X^-]E[Y^+] + E[X^-]E[Y^-]$$
$$= (E[X^+] - E[X^-])(E[Y^+] - E[Y^-])$$
$$= E[X]E[Y].$$

This establishes (7.10) for nonnegative random variables X and Y. ∎

In view of Theorem 7.13, we can define the notion of uncorrelated random variables.

Definition 7.3

We say that two real-valued random variables X and Y are *uncorrelated* if $E[XY] = E[X]E[Y]$.

The notion of uncorrelated random variables is closely related to the *covariance*, which is defined as

$$\mathrm{Cov}(X, Y) \triangleq E[(X - E[X])(Y - E[Y])].$$

It is straightforward to show that $\mathrm{Cov}(X, Y) = 0$ if and only if X and Y are uncorrelated. Theorem 7.13 says that two independent random variables are uncorrelated, but the converse does not hold in general.

Example 7.5.1 (Uncorrelated But Dependent Random Variables)
Let Z be any discrete random variable with zero mean. Let U be a random variable that equals 0 with probability 1/2 and 1 with probability 1/2. Define random variables X and Y by

$$(X, Y) = \begin{cases} (Z, Z) & \text{if } U = 0 \\ (Z, -Z) & \text{if } U = 1. \end{cases}$$

The random variable X has the same distribution as Z and hence has zero mean. The expectation $E[XY]$ is zero by a symmetry argument. Therefore X and Y are uncorrelated. Nevertheless, the random variables X and Y are not independent. It is because we can determine the absolute value of X if we know the value of Y.

Problems

7.1. Compute the mean and variance of the random variable specified by cdf

$$F(x) = \begin{cases} 0 & \text{if } x < 0 \\ x/3 & \text{if } 0 \leq x < 1 \\ 2/3 & \text{if } 1 \leq x < 2 \\ 1 & \text{if } 2 \leq x. \end{cases}$$

7.2 (Box–Muller Transform). To generate standard normal random variables, one can use a transformation involving two independent uniform random variables U and V, each of which takes values between 0 and 1. Specifically, generate a pair of random variables (X, Y) according to

$$X = \sqrt{-2 \log U} \cos(2\pi V),$$
$$Y = \sqrt{-2 \log U} \sin(2\pi V).$$

Show that this transformation induces the joint Gaussian distribution with mean $\mathbf{0}$ and covariance matrix $I_{2\times 2}$.

7.3. Let $f : [a, b] \to \mathbb{R}$ be a bounded function, $\alpha(x)$ a Stieltjes measure function on $[a, b]$, and μ the Lebesgue–Stieltjes measure induced by $\alpha(x)$. Assume $\mu(\{a\}) = 0$. Prove the following:

(a) f is Riemann–Stieltjes integrable on $[a, b]$ if and only if f is continuous μ-almost everywhere.

(b) If f is Riemann–Stieltjes integrable, then f is integrable with respect to the completion $\bar{\mu}$ of μ, and $\int_a^b f \, d\alpha = \int_{[a,b]} f \, d\bar{\mu}$.

7.4. Let X be an integrable function on a measure space $(\Omega, \mathcal{F}, \mu)$. Show that for any ϵ, there exists a set $E \in \mathcal{F}$ with $\mu(E) < \infty$ such that $\left| \int_{E^c} X \, d\mu \right| < \epsilon$.

7.5. Let $(X_i)_{i=1}^{\infty}$ be a sequence of nonnegative measurable functions defined on a common measure space $(\Omega, \mathcal{F}, \mu)$, such that $\int X_n \, d\mu < K$ for a finite constant K. Suppose that $(X_i)_{i=1}^{\infty}$ converges almost everywhere to a limit function X. Prove that X is integrable and $\int X \, d\mu \leq K$. (Hint: Apply Fatou's lemma.)

7.6 (Scheffé Lemma). Suppose f_n's are nonnegative measurable functions that converge to f almost everywhere as $n \to \infty$. Assume that $\int f_n \, d\mu \to \int f \, d\mu < \infty$.

(a) Show that $(f_k - f)^-$ is bounded by f, and hence, by the dominated convergence theorem, $\int (f_n - f)^- \to 0$.
(b) Show that $\int |f_n - f| \to 0$. (Hint: Write $|f_n - f|$ as $(f_n - f) + 2(f_n - f)^-$.)

7.7. Let $f_1, f_2, f_3, \ldots$ be measurable functions defined on a measure space $(\Omega, \mathcal{F}, \mu)$. Assume that $\sum_{k=1}^{\infty} \int |f_k| \, d\mu$ is finite.

(a) Show that $\sum_{k=1}^{\infty} f_k(\omega)$ converges to a finite limit μ-almost everywhere.
(b) Use part (a) to derive the following result:

$$\int \left(\sum_{k=1}^{\infty} f_k \right) d\mu = \sum_{k=1}^{\infty} \int f_k \, d\mu.$$

7.8. Let $(\Omega, \mathcal{F}, \mu)$ be a measure space, and consider a real-valued function $f : \Omega \times [a, b] \to \mathbb{R}$ satisfying the following conditions:

1. For any $x \in [a, b]$, the function $f(\omega, x)$ is Borel measurable and integrable with respect to μ.
2. There exists a set $E \in \mathcal{F}$ with $\mu(E) = 0$ such that for any $\omega \in E^c$, the derivative of the function $f(\omega, x)$ with respect to x exists and is equal to $f'(\omega, x)$.
3. There exists a function $g \in L^1(\mu)$ such that $|f'(\omega, x)| \leq g$ for μ-almost every ω and for all $x \in [a, b]$.

Prove that $f'(\omega, x)$ is integrable for any fixed x in $[a, b]$, and

$$\frac{d}{dx} \int_{\Omega} f(\omega, x) \, d\mu(\omega) = \int_{\Omega} f'(\omega, x) \, d\mu(\omega).$$

7.9. Let $(f_n)_{n \geq 1}$, $(g_n)_{n \geq 1}$, and $(h_n)_{n \geq 1}$ be sequences of integrable functions, satisfying the conditions:

1. $f_n \to f$, $g_n \to g$, and $h_n \to h$.
2. $f_n \leq g_n \leq h_n$, for all n.
3. $\int f_n \to \int f < \infty$ and $\int h_n \to \int h < \infty$.

Prove that $\int g$ is finite and $\int g_n \to \int g$.

Product Space and Coupling

8

Consider two real-valued continuous-type random variables X and Y with a joint probability density function $f_{XY}(x, y)$, defined on a common probability space. The probability of an event such as $\{\omega : X(\omega) \in B_1, Y(\omega) \in B_2\}$, where B_1 and B_2 are Borel sets, can be computed from the joint distribution. The marginal distributions of X and Y can also be derived from the joint distribution by integrating out the other variable.

Coupling is the reverse process of marginalization. We are given two random variables $\tilde{X}$ and $\tilde{Y}$ defined on separate probability spaces. The goal of coupling is to construct a new probability space that serves as a common source of randomness for random variables X and Y, and at the same time the law of X is the same as the law of $\tilde{X}$, and the law of Y is the same as the law of $\tilde{Y}$. A common coupling is the product space with product measure, in which the random variables X and Y are independent by construction. With the product measure, we can formulate the Tonelli–Fubini theorem, which is a fundamental theorem in the probability theory.

In general, the joint distribution of random variables X and Y need not be independent. When given a cost of $c(x, y)$ for moving from a point $x \in \mathcal{X}$ to a point y in $\mathcal{Y}$, we may look for a coupling that minimizes the expected cost. This optimization problem, called the optimal transport problem, arises in various fields including economics, statistics, and machine learning. In the last section in this chapter, we will derive the relationship between the total variation distance and maximal coupling.

8.1 Coupling

Coupling is a technique used to combine two or more stochastic experiments into a single experiment, enabling them to be studied and compared simultaneously. The key idea is to construct a common probability space that serves as a source of randomness for all the experiments. This new probability space is designed so that

© The Author(s), under exclusive license to Springer Nature Switzerland AG 2023
K. Shum, *Measure-Theoretic Probability*, Compact Textbooks in Mathematics,
https://doi.org/10.1007/978-3-031-49830-5_8

its marginal distributions match the given probability distributions of the individual experiment.

Definition 8.1

Given two probability spaces $(X, \mathscr{F}, P)$ and $(\mathcal{Y}, \mathscr{G}, Q)$, a *coupling* of P and Q consists of a common probability space $(\Omega, \mathscr{H}, \mu)$ and two measurable functions $X : \Omega \to X$ and $Y : \Omega \to \mathcal{Y}$ such that the push-forward measures $X_{\#}\mu$ and $Y_{\#}\mu$ are equal to P and Q, respectively.

There are several ways to construct a coupling between two probability distributions. One approach is to construct the product measure on $X \times \mathcal{Y}$, which makes X and Y independent. This will be the topic of the next section.

We may also consider constructing a transformation $T : X \to \mathcal{Y}$ such that $T(X)$ has the same distribution as Y. In this case, we can use the probability space for X as the common probability spaces for X and Y, resulting in a deterministic coupling between X and Y.

Definition 8.2

Continuing with the notation in Definition 8.1, if there exists a map $T : X \to \mathcal{Y}$ that is $(\mathscr{F}, \mathscr{G})$-measurable and satisfies $Y = T(X)$, we say that the coupling is *deterministic*. In this case, the function T is called a *transport map*.

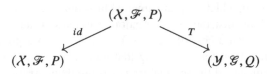

Deterministic coupling establishes a relationship between two random variables through a function. If the random variable X contains sufficient randomness to generate the random variable Y, then a coupling from X to Y exists.

Example 8.1.1 (Simulating a Bernoulli Random Variable by Tossing a Die)
We simulate a Bernoulli distributed random variable Y, with distribution $P(Y = 0) = 2/3$ and $P(Y = 1) = 1/3$, by tossing a fair die. Suppose X is a random variable that is uniformly distributed on $\{1, 2, 3, 4, 5, 6\}$. We can define Y to be 0 if $X = 1, 2, 3, 4$ and Y to be 1 if $X = 5, 6$. This function provides a deterministic coupling from X to Y.

To illustrate the concepts discussed above, we provide an example of two different couplings of Gaussian-distributed random variables below.

Example 8.1.2 (Couplings of Two Gaussian Random Variables)

Consider two normally distributed random variables $\tilde{X} \sim N(\mu_1, \sigma_1^2)$ and $\tilde{Y} \sim N(\mu_2, \sigma_2^2)$, where $\sigma_1 > 0$ and $\sigma_2 > 0$. There are several methods for constructing couplings between $\tilde{X}$ and $\tilde{Y}$.

The first method uses a deterministic coupling. We start by constructing a Lebesgue–Stieltjes measure P on $\mathbb{R}$ using the cumulative distribution function of $\tilde{X}$ as the Stieltjes measure function. We use the variable x to represent an element in this sample space. By construction, the identity function $X(x) = x$ is a random variable with the same distribution as $\tilde{X}$.

Next, we define the transformation

$$T(x) = \frac{\sigma_2}{\sigma_1}(x - \mu_1) + \mu_2. \tag{8.1}$$

Applying this transformation to X gives us $Y = T(X)$, which is a Gaussian random variable with the same distribution as $\tilde{Y}$. We note that in this coupling X and Y are certainly not independent, as one is a function of the other. Indeed, the joint distribution of X and Y is a singular Gaussian distribution.

The second method uses bivariate Gaussian distribution. We take the unit square $[0, 1] \times [0, 1]$ as the sample space Ω, with the Borel algebra on the unit square as the σ-algebra. Generate a random point p uniformly at random from the square, and let $U(p)$ and $V(p)$ denote the x- and y-coordinates of the point p, respectively. The random variables U and V are independent uniform random variables between 0 and 1. Next, we define $X = X(U(p), V(p))$ and $Y = Y(U(p), V(p))$ by the Box–Muller transformation (See Exercise 7.2). The resulting random variables $X \sim N(\mu_1, \sigma_1^2)$ and $Y \sim N(\mu_2, \sigma_2^2)$ are independent and have the same distribution as $\tilde{X}$ and $\tilde{Y}$, respectively.

Example 8.1.3 (Deterministic Coupling of Two Continuous Random Variables)

Suppose $\tilde{X}_1$ and $\tilde{X}_2$ are real-valued random variables with cdf $F_1(x)$ and $F_2(x)$, respectively. Furthermore, suppose that the two functions $F_1(x)$ and $F_2(x)$ are monotonically increasing functions from 0 to 1. For simplicity, suppose that the inverse functions of F_1 and F_2 exist. We can construct a coupling between $\tilde{X}_1$ and $\tilde{X}_2$ by defining X_1 and X_2 by $X_1 = F_1^{-1}(U)$ and $X_2 = F_2^{-1}(U)$, respectively, where U is a uniform random variable distributed between 0 and 1. (This is a preliminary idea of the Skorokhod representation theorem (Theorem 10.8).)

In what follows we investigate how to relate two couplings. Let $(\Omega, \mathcal{H}, \mu)$ and $(\Omega', \mathcal{H}, \mu')$ be couplings of probability measures P and Q, defined on the sample spaces $\mathcal{X}$ and $\mathcal{Y}$, respectively. We say that a measurable function $g : \Omega \to \Omega'$ is a *morphism* from the coupling on Ω to the coupling on Ω' if it satisfies the following commutative diagram:

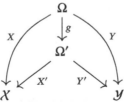

That is, we have $X = X' \circ g$ and $Y = Y' \circ g$. The measure μ' on Ω' is related to μ in Ω through the relation $\mu'(E') = \mu(g^{-1}(E')))$, which holds for all $E' \in \mathcal{H}$.

Using this notion of morphism, we can define the concept of *equivalent couplings*. Two couplings $(\Omega, \mathcal{H}, \mu)$ and $(\Omega', \mathcal{H}, \mu')$ are said to be equivalent if there exist two morphisms, g from the first coupling to the second and g' from the second to the first, that are inverses of each other up to a null set, meaning that the compositions $g' \circ g$ and $g \circ g'$ are equal to the identity map almost everywhere. To be precise, this means that for all $E \in \mathcal{H}$, we have $g'(g(E)) \triangle E = N$ and for all $E' \in \mathcal{H}'$, we have $g(g'(E')) \triangle E' = N'$, where N and N' are null sets in $\mathcal{H}$ and $\mathcal{H}'$, respectively, and $\triangle$ is the symmetric difference operator in set theory.

Suppose we have two separate sample spaces X and $\mathcal{Y}$. We want to turn the Cartesian product $X \times \mathcal{Y}$ into a measurable space by selecting a σ-algebra on $X \times \mathcal{Y}$. There are many ways to do this, but we are interested in the one that makes the two projection functions $\pi_1(x, y) = x$ from $X \times \mathcal{Y}$ to X and the projection $\pi_2(x, y) = y$ from $X \times \mathcal{Y}$ to $\mathcal{Y}$ measurable. Specifically, we want the inverse image $\pi_1^{-1}(E_1) = E_1 \times \mathcal{Y}$ to be measurable for all $E_1 \in \mathcal{F}$ and the inverse image $\pi_2^{-1}(E_2) = X \times E_2$ to be measurable for all $E_2 \in \mathcal{G}$.

This motivates the definition of measurable rectangles.

Definition 8.3

Let $(X, \mathcal{F})$ and $(\mathcal{Y}, \mathcal{G})$ be measurable spaces. For any $E_1 \in \mathcal{F}$ and $E_2 \in \mathcal{G}$, the Cartesian product $E_1 \times E_2$ is called a *measurable rectangle*. We use the notation $\mathcal{F} \times \mathcal{G}$ to represent the σ-field generated by all the measurable rectangles.

The σ-field $\mathcal{F} \times \mathcal{G}$ is the smallest σ-field containing all measurable rectangles. However, not all sets in $\mathcal{F} \times \mathcal{G}$ are of the form $E_1 \times E_2$ with $E_1 \in \mathcal{F}$ and $E_2 \in \mathcal{G}$. For example, the countable union of measurable rectangles is also in $\mathcal{F} \times \mathcal{G}$. Thus, there are many sets in $\mathcal{F} \times \mathcal{G}$ that are not themselves measurable rectangles.

The measurable space $(X \times \mathcal{Y}, \mathcal{F} \times \mathcal{G})$ serves as the basis on which we can build couplings. Furthermore, it can be shown that under certain condition, this construction is universal. The following theorem states that any coupling of $(X, \mathcal{F}, P)$ and $(\mathcal{Y}, \mathcal{G}, Q)$ can be factored through the product space $(X \times \mathcal{Y}, \mathcal{F} \times \mathcal{G})$.

Theorem 8.1

Suppose $(X, \mathcal{F})$, $(\mathcal{Y}, \mathcal{G})$, and $(\Omega, \mathcal{H})$ are measurable spaces in which all singletons are measurable. There is a unique measurable map

$$h : (\Omega, \mathcal{H}) \to (X \times \mathcal{Y}, \mathcal{F} \times \mathcal{G}),$$

which yields a morphism from any coupling $(\Omega, \mathcal{H}, \mu)$ of $(X, \mathcal{F}, P)$ and $(\mathcal{Y}, \mathcal{G}, Q)$ to $(X \times \mathcal{Y}, \mathcal{F} \times \mathcal{G}, h_{\#}\mu)$.

Proof We define the function $h : \Omega \to X \times \mathcal{Y}$ for any element $\omega \in \Omega$ as $h(\omega) = (X(\omega), Y(\omega))$. This function is measurable, meaning that for any measurable rectangle $E_1 \times E_2$ in $X \times \mathcal{Y}$, the inverse image $h^{-1}(E_1 \times E_2)$ is measurable. Specifically, $h^{-1}(E_1 \times E_2)$ can be expressed as the intersection of $X^{-1}(E_1)$ and $Y^{-1}(E_2)$, both of which are $\mathcal{H}$-measurable sets.

Suppose $(\Omega, \mathcal{H}, \mu)$ is a coupling of (P, Q). By definition, X is a measurable map from Ω to $\mathcal{X}$ such that $X_{\#}\mu = P$, and Y is a measurable map from Ω to $\mathcal{Y}$ such that $Y_{\#}\mu = Q$.

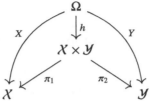

Consider the push-forward measure $h_{\#}\mu$ on $\mathcal{X} \times \mathcal{Y}$. By the definition of the functions π_1 and h, we have $\pi_1^{-1}(E_1) = E_1 \times \mathcal{Y}$ and

$$h^{-1}(E_1 \times \mathcal{Y}) = X^{-1}(E_1).$$

Therefore, for any $E_1 \in \mathcal{F}$, we have $(\pi_1 \circ h)^{-1}(E_1) = X^{-1}(E_1)$, and thus

$$\mu((\pi_1 \circ h)^{-1}(E_1)) = \mu(X^{-1}(E_1)) = P(E_1).$$

The last equality follows from the definition of coupling, which proves that the $(\mu_1 \circ h)_{\#}\mu = P$.

By a similar argument, we can show that $(\pi_2 \circ h)_{\#}\mu = Q$.

To show the uniqueness of h, we fix an element $x \in \mathcal{X}$ and an element $y \in \mathcal{Y}$. Since $\{x\}$ is a singleton, it is $\mathcal{F}$-measurable, and similarly, $\{y\}$ is $\mathcal{G}$-measurable. An element ω in the intersection of $X^{-1}(\{x\})$ and $Y^{-1}(\{y\})$ must be mapped to (x, y) in $\mathcal{X} \times \mathcal{Y}$. To see this, we can set P to be the Dirac measure concentrated at x and Q to be the Dirac measure concentrated at y (see Example 3.3.4). The Dirac measure μ that is concentrated at ω is a coupling of the two Dirac measures P and Q. In this case, defining $h(\omega)$ to be (x, y) is the only way to satisfy the requirements $(\pi_1 \circ h)_{\#}\mu = P$ and $(\pi_2 \circ h)_{\#}\mu = Q$, thereby establishing the uniqueness of the function h. ∎

As a corollary, we can restrict our attention to couplings defined on the product space without loss of generality. In terms of the product space, coupling of continuous random variables and coupling of discrete random variables can be reformulated as follows.

Definition 8.4

For two continuous random variables X and Y with pdf's $f_X(x)$ and $f_Y(x)$, respectively, a coupling of X and Y is a joint pdf $f_{XY} : \mathbb{R}^2 \to [0, \infty)$ such that $\iint_{\mathcal{X} \times \mathcal{Y}} f_{XY}(x, y)\, dx dy = 1$

$$f_X(x) = \int_{\mathcal{Y}} f_{XY}(x, y)\, dy, \quad \text{and } f_Y(y) = \int_{\mathcal{X}} f_{XY}(x, y)\, dx.$$

Similarly, for two discrete random variables X and Y with pmf's $p_X(x)$ and $p_Y(y)$, respectively, a coupling of X and Y is a joint pmf $p_{XY}(x, y)$ satisfying analogous properties.

8.2   Product Measure and Fubini Theorem

The product measure is a special coupling in which the random variables X and Y are independent. When X and Y are continuous random variables with given pdf's, we can simply define a coupling using the product of the two pdf's. In general, we can construct the product measure for any two σ-finite probability spaces $(\mathcal{X}_1, \mathcal{F}, P)$ and $(\mathcal{Y}, \mathcal{G}, Q)$, regardless of whether their distributions are continuous or discrete. The product measure is guaranteed to exist, and its construction is described in detail in standard probability theory textbooks, such as [2, Section 18] or [8, Chapter 8].

Theorem 8.2
Let $(\Omega_1, \mathcal{F}, P)$ and $(\Omega_2, \mathcal{G}, Q)$ be σ-finite measure spaces. Then there exists a unique measure, denoted by $P \times Q$, defined on the product space $(\mathcal{X} \times \mathcal{Y}, \mathcal{F} \times \mathcal{G})$, such that

$$(P \times Q)(E_1 \times E_2) = P(E_1)Q(E_2)$$

for all $E_1 \in \mathcal{F}$ and $E_2 \in \mathcal{G}$.

The proof of this theorem is omitted. A rough idea of the construction is to first define the product pre-measure on the measurable rectangles. Then we apply the measure extension theorem. The uniqueness follows by applying the π-λ theorem.

The utility of the product measure in probability theory is to combine two stochastic experiments into a single experiment, such that any event in the first experiment is independent from any event in the second one.

Example 8.2.1 (The Product of a Discrete and a Continuous Distribution)
Using Theorem 8.2, we are able to construct the product measure of two probability distributions of different types. Suppose $(\mathbb{R}, \mathscr{B}(\mathbb{R}), P)$ represents a Poisson distribution with mean λ, and $(\mathbb{R}, \mathscr{B}(\mathbb{R}), Q)$ represents Gaussian distribution with zero mean and unit variance. In this example, the product σ-algebra $\mathscr{B}(\mathbb{R}) \times \mathscr{B}(\mathbb{R})$ is the same as $\mathscr{B}(\mathbb{R}^2)$. Let $P \times Q$ be the product measure. We can calculate the probability of the event $\{n\} \times [a, b]$, where n is a positive integer and $a < b$ are real numbers,

$$(P \times Q)(\{n\} \times [a, b]) = \frac{\lambda^n}{n!}e^{-\lambda} \int_a^b \frac{1}{\sqrt{2\pi}}e^{-x^2/2}\, dx.$$

The probability in this example is concentrated on $\{0, 1, 2, \dots, \} \times \mathbb{R}$.

Once we have the product measure on the product space, we can compute the Lebesgue integral on it. One useful method for doing so is through iterative integration, which involves integrating with respect to one variable at a time. The Tonelli and Fubini theorems are mathematical tools that enable this approach. Before we state these two theorems, it is helpful to first consider a counter-example that motivates the need for these theorems.

Example 8.2.2 (A Counter-Example About Iterated Integrals)
Consider the double integral

$$\int_0^1 \int_0^1 \frac{x^2 - y^2}{(x^2 + y^2)^2} \, dx dy.$$

We claim that the two iterated integrals give different results. To see this, we use the fact that for any fixed constant a, the function $f(u) = \frac{a^2 - u^2}{(a^2 + u^2)^2}$ has anti-derivative $\frac{u}{a^2 + u^2} + C$, where C is a constant.

Suppose we integrate with respect to x first. To avoid division by zero at the origin, we integrate from ϵ to 1 in the y direction and take limit as ϵ tends to zero

$$\lim_{\epsilon \to 0} \int_\epsilon^1 \left(\int_0^1 \frac{x^2 - y^2}{(x^2 + y^2)^2} \, dx \right) dy = \lim_{\epsilon \to 0} \int_\epsilon^1 \left[\frac{-x}{x^2 + y^2} \right]_{x-0}^1 dy$$

$$= \lim_{\epsilon \to 0} \int_\epsilon^1 \frac{-1}{1 + y^2} \, dy = -\pi/4.$$

However, if we integrate with respect to y first, we will get the answer $\pi/4$.

The reason why the iterated integration in the previous example fails is because the function being integrated is not Lebesgue integrable. When the function to be integrated is nonnegative, the Tonelli theorem provides a condition under which the order integration can be exchanged, with possibility that the integral is infinite.

> **Theorem 8.3 (Tonelli Theorem)**
> If $f : X \times Y \to \bar{\mathbb{R}}$ is $(\mathcal{F} \times \mathcal{G})$-measurable and nonnegative and P and Q are σ-finite measures defined on X and Y, respectively, then
>
> $$\int_X \int_Y f(x, y) \, dQ(y) dP(x) = \int_Y \int_X f(x, y) \, dP(x) dQ(y)$$
>
> $$= \int_{X \times Y} f(x, y) \, d(P \times Q).$$

The first and second integrals in Theorem 8.3 are iterated integrals, while the last integral is an integral with respect to the product measure $P \times Q$. The equality in Theorem 8.3 is understood as follows: If one of the integrals is finite, then all integrals are finite and equal to one another; if one of them is infinite, then all of them are infinite. While we will not provide a proof of the Tonelli theorem here, we give an application of Tonelli theorem to the computation of the expected value of a nonnegative random variable (cf. Exercise 1.10).

Theorem 8.4

For any nonnegative random variable Y, we can compute the expectation $E[Y]$ by $\int_0^\infty P(Y \geq u)\, du$. If the cumulative distribution function is F_Y, we have $E[Y] = \int_0^\infty (1 - F_Y(u))\, du$.

Proof We write $\int_0^\infty P(Y \geq u)\, du$ as a double integral

$$\int_0^\infty P(Y \geq u)\, du = \int_{[0,\infty)} \int_\Omega \mathbf{1}_{\{y \geq u\}}\, dP(y) d\lambda(u),$$

where λ is the Lebesgue measure on $\mathbb{R}$. By the Tonelli theorem, we obtain

$$\int_0^\infty P(Y \geq u)\, du = \int_\Omega \int_{[0,\infty)} \mathbf{1}_{\{y \geq u\}}\, d\lambda(u) dP(y) = \int_\Omega y\, dP(y) = E[Y].$$

∎

If f is not nonnegative, we need to use the more general Fubini theorem to handle integration with product measure.

Theorem 8.5 (Fubini Theorem)

Let f be a real-valued or complex-valued measurable function defined on the product space $(\mathcal{X} \times \mathcal{Y}, \mathcal{F} \times \mathcal{G})$. Let P and Q be σ-finite measures defined on $\mathcal{X}$ and $\mathcal{Y}$, respectively. If one of the following integrals is finite:

$$\int_\mathcal{X} \int_\mathcal{Y} |f(x,y)|\, dQ(y) dP(x), \quad \int_\mathcal{Y} \int_\mathcal{X} |f(x,y)|\, dP(x) dQ(y),$$

$$\int_{\mathcal{X} \times \mathcal{Y}} |f(x,y)|\, d(P \times Q),$$

then $f \in L^1(P \times Q)$ and

$$\int_\mathcal{X} \int_\mathcal{Y} f(x,y)\, dQ(y) dP(x) = \int_\mathcal{Y} \int_\mathcal{X} f(x,y)\, dP(x) dQ(y)$$

$$= \int_{\mathcal{X} \times \mathcal{Y}} f(x,y)\, d(P \times Q).$$

The proof of this theorem can be found in textbooks on measure theory, such as [1] and [8].

In a typical application of Fubini theorem, we first take the absolute value of the function f and check if the iterated integral $\int_X \int_Y |f| \, dP \, dQ$ is finite using the Tonelli theorem. If it is finite, then we can apply the Fubini theorem to exchange the order of integration in $\int_X \int_Y f \, dP \, dQ$.

8.3 Application: Monge Problem and Kantorovich Problem

In this section, we will consider couplings of measures in which the sample space has more structure than a σ-field, such as a metric space, which allows us to define a notion of the distance between two points.

In 1781, the French mathematician Gaspard Monge proposed an optimization problem now known as the Monge transport problem. The problem involves moving a pile of sands to fill up a hole with the same volume using minimum effort. The cost of moving a unit of sand is measured by the distance traveled.

To formulate this problem mathematically, we view the distribution of sand as a measure μ on $\mathbb{R}^2$ and the shape of the hole as another measure ν on $\mathbb{R}^2$. We define the *support* of a measure μ as the smallest closed subset in which every open neighborhood of any point of the set has positive μ-measure. We denote this set by $\text{supp}(\mu)$. The assumption that the hole has the same volume as the sand pile is translated to

$$\mu(\text{supp}(\mu)) = \nu(\text{supp}(\nu)).$$

A *transport map* is a function T from $\text{supp}(\mu)$ to $\text{supp}(\nu)$ such that the image measure $T_{\#}\mu$ is the same as measure ν. The objective is to find the optimal transport map T that minimizes the total cost

$$\int_{\text{supp}(\mu)} |\omega - T(\omega)| \, d\mu(\omega)$$

over all transport maps T such that $T_{\#}\mu = \nu$. This problem can be generalized by considering a more general cost function $c(x, y)$. Despite its generality, the Monge transport problem is highly nonlinear, and finding the optimal solution can be challenging.

We illustrate a few special cases below. For notation simplicity, we consider one-dimensional space $\mathbb{R}$ and take the squared distance $(x - y)^2$ as the distance measure $c(x, y)$.

Example 8.3.1 (A Degenerate Example with Zero Transport Cost)
Suppose random variables $\tilde{X} \sim N(0, \sigma^2)$ and $\tilde{Y} \sim N(0, \sigma^2)$ have the same distribution. Then the obvious transport map is the identity map, which actually does nothing. The optimal transport cost is 0.

Example 8.3.2 (A Finite Version of Monge's Problem)
Let δ_x denote the Dirac measure of a point $x \in \mathbb{R}$ (See Example 3.3.4). Suppose there is one unit
of sand located at point x_i, for $i = 1, 2, \ldots, n$. The input measure μ is

$$\mu = \sum_{i=1}^{n} \delta_{x_i}.$$

We want to move the sands to n other points $y_i, i = 1, 2, \ldots, n$, such that each point y_i receives
exactly one unit of sand. The output measure ν is thus

$$\nu = \sum_{i=1}^{n} \delta_{y_i}.$$

With this data, Monge's problem becomes a combinatorial problem, which is to find the optimal
permutation $\pi : \{1, 2, \ldots, n\} \rightarrow \{1, 2, \ldots, n\}$ that minimizes the transportation cost given by

$$\sum_{i=1}^{n} (x_i - y_{\pi(i)})^2.$$

In other words, we want to find the optimal way to move each unit of sand from its initial
position x_i to its final position $y_{\pi(i)}$ such that the total transport cost is minimized.

Example 8.3.3 (Nonexistence of Solution in Monge's Problem)
Monge's problem may fail to have a solution, because it is not allowed to split any mass. Consider
the measures μ and ν given by $\mu = 2\delta_{x_1}$ and $\nu = \delta_{y_1} + \delta_{y_2}$, where x_1, y_1, and y_2 are distinct
real numbers. The measure μ represents two units of sand at position x_1, and we want to move
the sand to positions y_1 and y_2. Suppose that the three points x_1, y_1, and t_2 are distinct. There is
no transport map T such that $T_{\#}\mu = \nu$, because any transport map must move both units of sand
from x_1 to either y_1 or y_2, which violates the constraint that we cannot split the mass. Therefore,
Monge's problem does not have a solution in this case.

Kantorovich's problem can be seen as a relaxation of Monge's problem by
allowing the splitting of mass. One of the advantages of the Kantorovich problem
is that it guarantees the existence of feasible solution, even when Monge's problem
does not have a solution. Moreover, in finite sample spaces, the optimal solution can
be found efficiently using linear programming.

The Kantorovich version of the transport problem differs from Monge's problem
in that it deals with transport plan rather than transport map. To illustrate this, we
consider an example borrowed from Villani [9]. There are m bread bakeries in a
city that produce bread and n restaurants that demand bread on a daily basis. The
amount of bread produced by each bakery and the demand of each restaurant are
known in advance. The optimization problem is to minimize the total transport cost
of bread from the bakeries to the restaurants.

In contrast to Monge's problem, the output of a bakery can be sent to different
restaurants. This is possible because the unit of bread is assumed to be infinitely
divisible. The problem can be interpreted as finding the optimal coupling of the
production measure (which describes the amount of bread produced by each bakery)

and the demand measure (which describes the amount of bread demanded by each restaurant).

Before we formally state the Kantorovich problem, we first define some relevant concepts. For simplicity, let X and Y be metric spaces, and let $\mathcal{B}(X \times Y)$ be the Borel algebra that is generated by the open sets in the form $A \times B$, where A and B are open sets in X and Y, respectively. We define the projection function π_1 from $X \times Y$ to X by $\pi_1(x, y) = x$ and the projection to Y by $\pi_2(x, y) = y$. The *marginals* of a probability measure μ on $X \times Y$ are the image measures $(\pi_1)_{\#}\mu$ and $(\pi_2)_{\#}\mu$, which are probability measures on X and Y, respectively. A *transport plan* between two probability measures P and Q defined on X and Y, respectively, is a probability measure on $X \times Y$ with marginals P and Q.

Let $\mathcal{PM}(X \times Y)$ denote the set of all probability measures on the measurable space $(X \times Y, \mathcal{B}(X \times Y))$. Given a cost function $c(x, y) : X \times Y \to \mathbb{R}_{\geq 0}$, the Kantorovich problem is to find a transport plan μ in

$$\Pi(P, Q) \triangleq \{\mu \in \mathcal{PM}(X \times Y) : (\pi_1)_{\#}\mu = P, \ (\pi_2)_{\#}\nu = Q\},$$

which minimizes

$$E[c(X, Y)] = \int_{X \times Y} c(x, y) \, d\mu(x, y). \tag{8.2}$$

An optimal transport plan μ^* is a transport plan that solves the Kantorovich problem with the minimum cost.

The spaces X and Y in this optimization problem can be finite sets or closed intervals in $\mathbb{R}$. We give a discrete example below.

Example 8.3.4 (A Discrete Kantorovich Problem)
Suppose $X = \{a, b\}$ is a set with size 2, and $Y = \{1, 2, 3\}$ is a set with size 3. Let P be the probability measure on X defined by $P(\{a\}) = 1/3$ and $P(\{b\}) = 2/3$, and let Q be the probability measure on Y defined by $Q(\{1\}) = Q(\{2\}) = 1/5$ and $Q(\{3\}) = 3/5$. The Kantorovich problem is to find a transport plan between P and Q that minimizes the total transport cost, where the cost of transporting mass from $i \in X$ to $j \in Y$ is given by the cost function $c : X \times Y \to \mathbb{R}$.

The transport plan can be represented by a 2×3 matrix $T = [t_{ij}]$, where t_{ij} represents the amount of mass transported from $i \in X$ to $j \in Y$. The cost of transporting mass from i to j is given by $c_{ij} = c(i, j)$. The Kantorovich problem can be formulated as a linear programming problem with variables t_{ij}. The objective function to be minimized is

$$\sum_{i \in \{a, b\}} \sum_{j \in \{1, 2, 3\}} c_{ij} t_{ij}.$$

The constraints are the marginal constraints that ensure that the total mass transported from each source equals the total mass received at each target:

$$t_{a1} + t_{b1} = 1/5, \quad t_{a1} + t_{a2} + z_{a3} = 1/3,$$

$$t_{a2} + t_{b2} = 1/5, \quad t_{b1} + t_{b2} + z_{b3} = 2/3,$$

$$t_{a3} + t_{b3} = 3/5,$$

$$t_{ij} \geq 0 \quad \text{for } i \in \{a, b\}, j \in \{1, 2, 3\}.$$

Solving this linear programming problem gives the optimal transport plan between P and Q.

In the Kantorovich problem, if X is the same as Y and is equipped with a metric $d(x, y)$, we can take $d(x, y)^p$ as the cost function, for some real number $p \geq 1$. The p-th root of the minimal transport cost is called the p-*Wasserstein distance* between P and Q

$$W_p(P, Q) \triangleq \left(\min_{\mu \in \Pi(P,Q)} \int_{X \times X} d(x, y)^p d\mu(x, y) \right)^{1/p}.$$

The optimal transport problem with Wasserstein distance is an emerging method in computer vision, statistics, and data science [6].

8.4   Application: Total Variation Distance

In statistics and the theory of stochastic processes, it is often important to determine whether two probability distributions are similar to each other and to provide a quantitative measure of the distance between them. This is particularly useful in areas such as hypothesis testing and model selection.

A good distance measure should capture the basic features of a distance, i.e., it is a metric function. Let M be a collection of probability measures defined on a measurable space. We want a function $d : M \times M \to \mathbb{R}$ satisfying the following properties, known as the metric properties:

- (Nonnegativity) $d(P, Q) \geq 0$, with equality if and only if $P = Q$.
- (Symmetry) $d(P, Q) = d(Q, P)$ for any $P, Q \in M$.
- (Triangle inequality) $d(P, R) \leq d(P, Q) + d(Q, R)$ for any $P, Q, R \in M$.

There are many distance measures in the probability theory that satisfy the metric properties, and the choice of distance measure depends on the specific problem at hand. For example, the Wasserstein distance mentioned at the end of the last section is indeed a distance function. In this section, we will consider the total variation (TV) distance.

Definition 8.5

The *total variation distance* of two probability measures P and Q on a common measurable space $(\Omega, \mathcal{F})$ is defined as

$$d_{TV}(P, Q) \triangleq \sup_{A \in \mathcal{F}} |P(A) - Q(A)|. \tag{8.3}$$

In other words, the total variation distance measures the largest possible difference between the probabilities that the two measures assign to the same event, and it is always bounded between 0 and 1. As we will see later, the total variation distance is closely related to the concept of coupling.

Example 8.4.1 (Total Variation Distance Between Two Dirac Measures)

Let μ_x denote the Dirac measure on $\mathbb{R}$ that is concentrated at the point x, i.e., $\mu_x(A) = 1$ if $x \in A$ and $\mu_x(A) = 0$ if $x \notin A$ (See Example 3.3.4). For $x \neq y$, the total variation distance between μ_x and μ_y is equal to 1. It is because for any measurable set A that contains x but not y, we have $\mu_x(A) - \mu_y(A) = 1 - 0 = 1$.

Example 8.4.2 (Total Variation Distance Between Two Bernoulli Distributions)

Let $\Omega = \{H, T\}$ be a sample space with two outcomes. We can define two probability measures P and Q on Ω by setting $P(\{H\}) = p = 1 - P(\{T\})$ and setting $Q(\{H\}) = q = 1 - Q(\{T\})$, where $0 \leq p, q \leq 1$. If we take $A = \{H\}$ in (8.3), we obtain $|P(\{H\}) - Q(\{H\})| = |p - q|$. Likewise, when we take $A = \{T\}$, we obtain $|P(\{T\}) - Q(\{T\})| = |p - q|$. The total variation distance is $|p - q|$.

When $p = q$, the two probability measures P and Q are the same, and their total variation distance is 0. Otherwise, the total variation distance is the absolute difference between the two probabilities p and q.

When two probability distributions are both of discrete type and both continuous type, we can compute the total variation distance by the formulas in the following theorem.

Theorem 8.6

Suppose P and Q are probability measures defined on the set of nonnegative integers $\Omega = \{0, 1, 2, 3, \ldots\}$. Then, the total variation distance between P and Q is given by

$$d_{TV}(P, Q) = \frac{1}{2} \sum_{i=0}^{\infty} |P(\{i\}) - Q(\{i\})|.$$

If probability measures P and Q have piece-wise continuous pdf's $f(x)$ and $g(x)$, respectively, then we have

$$d_{TV}(P, Q) = \frac{1}{2} \int_{-\infty}^{\infty} |f(x) - g(x)| \, dx.$$

Proof We only prove the discrete case. We first note that in the definition of total variation distance, we can remove the absolute value in (8.3) without changing the result. This is because if $P(A) < Q(A)$, we can consider the complement A^c and use the fact that

$$|P(A) - Q(A)| = |1 - P(A^c) - (1 - Q(A^c))| = P(A^c) - Q(A^c).$$

Hence, the computation of total variation distance amounts to the maximization of $P(A) - Q(A)$ over all events A. We claim that is maximized when A is the event

$$A_+ \triangleq \{i \in \Omega : P(\{i\}) > Q(\{i\})\}.$$

To see this, we let $A_- \triangleq \{i : P(\{i\}) < Q(\{i\})\}$, and $A_0 = \{i : P(\{i\}) = Q(\{i\})\}$. The three events A_+, A_-, and A_0 are disjoint. Hence,

$$P(A_+) + P(A_-) + P(A_0) = 1 = Q(A_+) + Q(A_-) + Q(A_0).$$

By noting $P(A_0) = Q(A_0)$, we obtain

$$|P(A_+) - Q(A_+)| = |P(A_-) - Q(A_-)| = \frac{1}{2} \sum_{i=0}^{\infty} |P(\{i\}) - Q(\{i\})|.$$

The proof for the continuous case is similar.   ∎

Example 8.4.3 (Total Variation Distance Between Bernoulli and Poisson)
Let P be a Bernoulli distribution with $P(\{1\}) = \lambda$ and $P(\{0\}) = 1 - \lambda$ and Q be a Poisson distribution with mean λ, where λ is a real number between 0 and 1. For $k \geq 0$, $Q(\{k\}) = \frac{\lambda^k}{k!} e^{-\lambda}$. Applying Theorem 8.6, we obtain

$$d_{TV}(P, Q) = \frac{1}{2}|1 - \lambda - e^{-\lambda}| + \frac{1}{2}|\lambda - \lambda e^{-\lambda}| + \frac{1}{2} \sum_{k=2}^{\infty} \frac{\lambda^k}{k!} e^{-\lambda}.$$

The infinite sum on the right-hand side can be simplified to $\frac{1}{2}(1 - e^{\lambda} - \lambda e^{-\lambda})$. We thus have

$$d_{TV}(P, Q) = \frac{1}{2}|1 - \lambda - e^{-\lambda}| + \frac{1}{2}|\lambda - \lambda e^{-\lambda}| + \frac{1}{2}(1 - e^{\lambda} - \lambda e^{-\lambda}).$$

Since $1 - \lambda$ is less than or equal to $e^{-\lambda}$ for all λ, we can simplify this expression as follows:

$$d_{TV}(P, Q) = \frac{1}{2}\left[e^{-\lambda} + \lambda - 1 + \lambda - \lambda e^{-\lambda} + 1 - e^{\lambda} - \lambda e^{-\lambda}\right] = \lambda(1 - e^{-\lambda}).$$

Note that when $\lambda = 0$, the Bernoulli distribution with mean 0 and the Poisson distribution with mean 0 are the same distribution. As λ increases from 0 to 1, the total variation distance increases, indicating that the distributions become increasingly different as λ increases.

In the rest of this section, we assume that the sample space $\mathcal{X}$ is a Polish space, i.e., it is separable and equipped with a complete metric function. To be precise, we assume that the space $\mathcal{X}$ has a countable dense subset, and we can measure the distance between two points in $\mathcal{X}$ using a complete metric. We take the Borel algebra generated by the open sets in $\mathcal{X}$ as the σ-algebra $\mathcal{F}$. This is a convenient and natural assumption in the study of total variation distance and optimal transport problems in general. A measurable space $(\mathcal{X}, \mathcal{F})$ arising in this way is also called a *standard Borel space*. It is worth noting that any discrete space and Euclidean space $\mathbb{R}^d$ are Polish. Hence, discrete random variables and continuous random variables can all be defined on Polish spaces.

It is interesting to construct coupling such that the probability of the event $X = Y$ is as large as possible. This is the diagonal event $\{(\omega, \omega) : \omega \in \mathcal{X}\}$ in the product space $\mathcal{X} \times \mathcal{X}$. With the additional assumptions stated in the previous paragraph, we can guarantee that the diagonal event in the product space is measurable, so that the event $\{X = Y\}$ has well-defined probability. The coupling inequality below gives a lower bound on the extent of how much we can make X equal to Y.

Theorem 8.7 (Coupling Inequality)
Given two probability measures P and Q defined on the same measurable space $(\mathcal{X}, \mathcal{F})$, any coupling of (X, Y) defined on a probability space $(\Omega, \mathcal{H}, \mu)$ satisfies

$$d_{TV}(P, Q) \leq \mu(\{X \neq Y\}).$$

Proof The proof depends on the following trick. For any subset $A \subseteq \mathcal{X}$ that is $\mathcal{F}$-measurable, we have

$$
\begin{aligned}
P(A) - Q(A) &= \mu(X \in A) - \mu(Y \in A) \\
&= \mu(X \in A, X = Y) + \mu(X \in A, X \neq Y) \\
&\quad - \mu(Y \in A, X = Y) - \mu(Y \in A, X \neq Y) \\
&= \mu(X \in A, X \neq Y) - \mu(Y \in A, X \neq Y) \\
&\leq \mu(X \in A, X \neq Y).
\end{aligned}
$$

Taking supremum over all $A \in \mathcal{F}$ on both sides, we get

$$\sup_{A \in \mathcal{F}} (P(A) - Q(A)) \leq \sup_{A \in \mathcal{F}} \mu(X \in A, X \neq Y).$$

The last supremum is achieved when $A = \mathcal{X}$. This proves $d_{TV}(P, Q) \leq \mu(X \neq Y)$.
∎

A coupling (X, Y) that attains the coupling inequality is called a *maximal coupling*. It can be shown that maximal coupling always exists when the measurable space $(\mathcal{X}, \mathcal{F})$ is Polish [9, Theorem 4.1].

Example 8.4.4 (An Example of Maximal Coupling)
Suppose $\Omega = \{1, 2\}$ and we define two probability measures P and Q on X by

$$P(\{1\}) = 1/3, \;\; P(\{2\}) = 2/3, \qquad \text{and } \; Q(\{1\}) = 3/4, \;\; P(\{2\}) = 1/4.$$

The total variation distance between P and Q is $2/3 - 1/4 = 5/12$.
The corresponding Kantorovich problem is to minimize $x_{12} + x_{21}$, subject to the constraints

$$x_{11} + x_{12} = 1/3, \qquad x_{11} + x_{21} = 3/4,$$

$$x_{21} + x_{22} = 2/3, \qquad x_{12} + x_{22} = 1/4,$$

$$x_{11}, x_{12}, x_{21}, x_{22} \geq 0.$$

We want to take the variables x_{12} and x_{21} to be as small as possible. In other words, we want x_{11} and x_{22} to be as large as possible, while satisfying the constraints. Because all variables are nonnegative, we must have $x_{11} \leq 1/3$ and $x_{22} \leq 1/4$. If we set $x_{11} = 1/3$ and $x_{22} = 1/4$, then the other constraints imply that we must have $x_{12} = 0$ and $x_{21} = 2/3 - 1/4 = 5/12$. We can visualize the transport plan as a 2×2 array

$$\begin{bmatrix} x_{11} & x_{12} \\ x_{21} & x_{22} \end{bmatrix} = \begin{bmatrix} 1/3 & 0 \\ 5/12 & 1/4 \end{bmatrix}.$$

This is indeed the optimal solution to the linear program. It agrees with the total variation distance $5/12$ of P and Q and hence is a maximal coupling.

The total variation distance can be interpreted as a solution to the Kantorovich problem with cost function

$$c_0(x, y) = \begin{cases} 1 & \text{if } x \neq y \\ 0 & \text{if } x = y. \end{cases}$$

For any coupling $(X \times X, \mathscr{F} \times \mathscr{F}, \mu)$ of P and Q, the probability $\mu(\{X \neq Y\})$ can be written as

$$\int_{X \times X} c_0(x, y) \, d\mu(x, y).$$

Let $\Pi(P, Q)$ denote the set of all probability measures μ on the product space $(X \times X, \mathscr{F} \times \mathscr{F})$ such that $X_{\#}\mu = P$ and $Y_{\#}\mu = Q$. The coupling inequality implies that

$$d_{TV}(P, Q) = \inf_{\mu \in \Pi(P, Q)} \left\{ \int_{X \times X} c_0(x, y) d\mu(x, y) \right\}. \tag{8.4}$$

The infimum is taken over all probability measures $\mu \in \Pi(P, Q)$, and when the measurable space $(X, \mathscr{F})$ is Polish, it can be achieved by a maximal coupling. The existence of a maximal coupling illustrates the close connection between the total variation distance and coupling.

Problems

8.1. Let X and Y be the set of natural numbers $\mathbb{N}$ and the σ-algebras the power set $\mathcal{P}(\mathbb{N})$. Let μ and ν be the counting measures on X and Y, respectively. We define a function $f : X \times Y \to \mathbb{R}$ by

$$
f(i, j) \triangleq \begin{cases} i & \text{if } i = j \\ -i & \text{if } j = i + 1 \\ 0 & \text{otherwise.} \end{cases}
$$

Show that the conclusion of the Fubini theorem fails in this example.

8.2. Apply Fubini's theorem to prove that the order of summation in $\sum_{i=0}^{\infty} \sum_{j=0}^{\infty} a_{ij}$ is irrelevant if $\sum_{i=0}^{\infty} \sum_{j=0}^{\infty} |a_{ij}|$ is finite.

8.3. Let P and Q be exponential distributions with parameters λ and μ, respectively. Find the total variation distance between P and Q.

8.4. Define two probability distributions P and Q on a finite sample space $\Omega = \{a, b, c\}$. The pmf's of P and Q are given in the following table:

ω	a	b	c
$P(\omega)$	1/3	1/3	1/3
$Q(\omega)$	1/2	1/4	1/4

Find the maximal coupling of P and Q.

8.5. Let μ and ν be continuous-type probability measures with pdf $f(x)$ and $g(x)$, respectively. Show that for any bounded measurable function h and Borel set B,

$$
\left| \int_B h(x) f(x)\, dx - \int_B h(x) g(x)\, dx \right| \le 2 d_{TV}(\mu, \nu) \cdot \sup_{x \in B} h(x).
$$

8.6 (Poisson Approximation of Binomial Distribution). Suppose $X = X_1 + X_2 + \cdots + X_n$ and $Y = Y_1 + Y_2 + \cdots + Y_n$ are sums of random variables. We use the notation Ber(λ) for Bernoulli distribution with mean λ and Poi(λ) for Poisson distribution with mean λ.

(a) Show that $P(X \ne Y) \le \sum_{k=1}^{n} P(X_k \ne Y_k)$. (Show that the event $\{X \ne Y\}$ is a subset of $\cup_{k=1}^{n} \{X_k \ne Y_k\}$.)
(b) Construct a maximal coupling of Ber(λ) and Poi(λ).

(c) For $k = 1, 2, \ldots, n$, let $\hat{X}_k$ and $\hat{Y}_k$ be the maximal coupling of Ber(λ_k) and Poi(λ_k) obtained in part (b). Show that the total variation between $\hat{X} = \hat{X}_1 + \cdots + \hat{X}_n$ and $\hat{Y} = \hat{Y}_1 + \cdots + \hat{Y}_n$ is upper bounded by $\sum_{k=1}^{n} \lambda_k^2$.

(d) Derive an upper bound of total variation distance between Binom($n, \lambda/n$) and Poi(λ) by substituting $\lambda_k = \lambda/n$ in part (c).

8.7. Let X and Y be continuous-type random variables with pdf $f_X(x)$ and $f_Y(y)$, respectively. Suppose both $f_X(x)$ and $f_Y(y)$ are piece-wise continuous. We construct random variables $\hat{X}$ and $\hat{Y}$ as follows. First compute $p \triangleq \frac{1}{2} \int_{-\infty}^{\infty} |f_X(t) - f_Y(t)| \, dt$. Let B be a Bernoulli random variable with success probability p. Define three more auxiliary random variables U, V, and W with the following pdf's:

$$f_U(u) = \min(f_X(u), f_Y(u))/(1 - p)$$

$$f_V(v) = (f_X(v) - \min(f_X(v), f_Y(v)))/p$$

$$f_W(w) = (f_Y(w) - \min(f_X(w), f_Y(w)))/p.$$

Generate random variables B, U, V, and W such that they are mutually independent. Define $(\hat{X}, \hat{Y}) := (U, U)$ if $B = 0$ and $(\hat{X}, \hat{Y}) := (V, W)$ if $B = 1$. Prove that this gives a maximum coupling of the pdf f_X and f_Y.

Moment Generating Functions and Characteristic Functions

9

The moment generating function and characteristic function are both examples of transform functions used to analyze probability distribution. The moment generating function of a random variables contains all the information about its moments. However, a drawback of the moment generating function is that it may not exist in certain cases. In this chapter we will prove that the moment generating function of a random variable exists if and only if the moments of all orders are finite.

In contrast, the characteristic function of a random variable is well-defined for all types of random variables, even if the moments are not finite. The inversion formula and the uniqueness theorem state that we can recover not only the moments, but the entire probability distribution of a random variable from its characteristic function. Although the inverse formula is not frequently used to compute probability explicitly, it provides a theoretical basis for checking convergence in distribution through characteristic functions.

9.1   Moments and Moment Generating Functions

The moments provides some information about the shape of the probability distribution.

Definition 9.1

For integer $r \geq 1$, the *r-th moment* of X is defined as the expectation $E[X^r]$. The *r-th central moment* is defined by $E[(X - E[X])^r]$. In particular, the second central moment is commonly called the *variance* of X; the square root of variance is called the *standard deviation*.

The third central moment measures the asymmetry of the probability distribution. The *skewness* of a random variable is defined as the third central moment normalized by the cube of the standard deviation σ,

© The Author(s), under exclusive license to Springer Nature Switzerland AG 2023   149
K. Shum, *Measure-Theoretic Probability*, Compact Textbooks in Mathematics,
https://doi.org/10.1007/978-3-031-49830-5_9

$$\frac{E[(X - E[X])^3]}{\sigma^3}.$$

The analogous quantity of order 4 is called the *kurtosis*,

$$\frac{E[(X - E[X])^4]}{\sigma^4}.$$

It measures the "tailedness" of the probability distribution.

By Theorem 7.12, we have the following formula for computing the moments of continuous-type random variables.

Theorem 9.1

Suppose X is a random variable such that $E[|X|^r] < \infty$. If the cdf $F_X(x)$ of X is differentiable and the derivative equals $f_X(x)$, we can compute the r-th moment and the r-th central moment by the following formulas:

$$E[X^r] = \int_{-\infty}^{\infty} x^r f_X(x)\, dx,$$

$$E[(X - E[X])^r] = \int_{-\infty}^{\infty} (x - E[X])^r f_X(x)\, dx.$$

To compute the higher-order moments, we can use the moment generating function.

Definition 9.2

Let X be a real-valued random variable defined on a probability space $(\Omega, \mathscr{F}, P)$. The *moment generating function* (mgf) of X is defined as

$$M_X(t) \triangleq E[e^{tX}] = \int_{\Omega} e^{tX(\omega)}\, dP(\omega),$$

where $t \in \mathbb{R}$. We say that X *has a moment generating function* if there exists $\delta > 0$ such that $M_X(t)$ is finite for all t in the interval $(-\delta, \delta)$.

Since X is real-valued, the function e^{tX} is nonnegative, and hence the expectation $E[e^{tX}]$ is well-defined. However, the expected value may be infinity. We note that it is essential to have $M_X(t)$ defined and finite in a non-empty interval $(-\delta, \delta)$ that contains 0 in the interior. This ensures that we can differentiate $M_X(t)$ with respect to t at $t = 0$.

> **Theorem 9.2**
> *If X has moment generating function $M_X(t)$, then X has finite moments of all orders, and they can be recovered from the moment generating function by*
>
> $$E[X^n] = \frac{d^n}{dt^n} M_X(t)\Big|_{t=0}$$
>
> *for $n = 1, 2, 3, \ldots$*

Proof In this proof we will use the inequality

$$e^{|tx|} \leq e^{tx} + e^{-tx},$$

which holds for any real numbers t and any x. By assumption, there exists $\delta > 0$ such that $M_X(t)$ is finite for $-\delta < t < \delta$. Hence, $E[e^{tX}] < \infty$ and $E[e^{-tX}] < \infty$ for $|t| < \delta$.

We first prove that all moments of X are finite using Taylor expansion. For any positive integer n, we have

$$\frac{|t|^n}{n!} |x|^n \leq e^{|tx|},$$

because the left-hand side is just one of the terms in the power series expansion of $e^{|tx|}$, and all other terms in the expansion are positive. Substituting t by $\delta/2$, we obtain

$$\frac{\left(\frac{\delta}{2}\right)^n}{n!} |x|^n \leq e^{\delta|x|/2}.$$

Since this holds for all X, we can replace x by random variable X and take expectation on both sides. This gives

$$\frac{\left(\frac{\delta}{2}\right)^n}{n!} E[|X|^n] \leq E[e^{\delta|X|/2}]$$

$$E[|X|^n] \leq \frac{n!}{\left(\frac{\delta}{2}\right)^n} \left[E[e^{\delta X/2}] + E[e^{-\delta X/2}] \right] < \infty.$$

Hence, by Theorem 6.6, the random variable X^n is integrable for all n.

Next, from the Taylor expansion of the exponential function, we can derive the following inequality that holds for any real numbers t and x, and for all positive integers n,

$$\left|\sum_{k=0}^{n} \frac{t^k}{k!} x^k\right| \leq \sum_{k=0}^{n} \frac{|t|^k}{k!} |x|^k \leq e^{|t||x|}.$$

For each t in the range $|t| < \delta$, the sum $\sum_{k=0}^{n} t^k X^k / k!$ is thus bounded by $e^{|tX|}$. On the other hand, $e^{|tX|}$ has finite expectation, because

$$E[e^{|t||X|}] \leq E[e^{tX}] + E[e^{-tX}] < \infty$$

for all $|t| < \delta$. We can apply the dominated convergence theorem (Theorem 7.4) to obtain

$$E[e^{tX}] = E\left[\sum_{k=0}^{\infty} \frac{t^k X^k}{k!}\right] \overset{\text{DCT}}{=} \sum_{i=0}^{\infty} E\left[\frac{t^i X^i}{i!}\right] = \sum_{i=0}^{\infty} \frac{E[X^i]}{i!} t^i.$$

Therefore, $M_X(t)$ can be expanded as a power series with radius of convergence δ. Using the properties of power series, we see that $M_X(t)$ is infinitely differentiable and

$$\left.\frac{d^n}{dt^n} M_X(t)\right|_{t=0} = E[X^n].$$

∎

▶ **Existence of Moment Generating Function** Theorem 9.2 illustrates that a random variable need not have a moment generating function. A necessary condition for the existence is that all moments are finite. For example, if a random variable has infinite variance, then it does not have a moment generating function. Moreover, it is possible that even if all moments are finite, the moment generating function still does not exist. One such example is the log-normal distribution.

Example 9.1.1 (Moment Generating Function of Poisson Distribution)
Suppose X is a Poisson random variable with mean λ. The probability mass function is $P(X = k) = \frac{\lambda^k}{k!} e^{-\lambda}$, for $k = 0, 1, 2, \ldots$. Using the discrete version of the change-of-variable formula (Theorem 7.12), we can calculate the moment generating function as

$$M_X(t) \triangleq E[e^{tX}] = \sum_{k=0}^{\infty} e^{kt} \cdot \frac{\lambda^k}{k!} e^{-\lambda} = e^{-\lambda} \sum_{i=0}^{\infty} (\lambda e^t)^k / k! = e^{\lambda e^t - \lambda}.$$

$M_X(t)$ is defined for all t.
The first two moments of X can be obtained by differentiating $M_X(t)$,

$$E[X] = M_X'(0) = e^{-\lambda}(e^{\lambda e^t} \cdot \lambda e^t)\Big|_{t=0} = \lambda$$

$$E[X^2] = \frac{d^2}{dt^2} M_X(0) = e^{-\lambda}(e^{\lambda e^t}(\lambda e^t)(\lambda e^t) + e^{\lambda e^t} \lambda e^t)\Big|_{t=0} = \lambda^2 + \lambda.$$

The variance of X is thus $E[X^2] - E[X]^2 = \lambda$.

9.2   Characteristic Functions

The main disadvantage of moment generating function is that it does not exist for some random variables. A better option is the characteristic function, which is similar to the moment generating function but requires complex function and complex integration (See Definition 6.6). The characteristic function can be regarded as the Fourier transform of the probability distribution function.

The characteristic function is an essential tool in determining whether a sequence of random variables converges in distribution. We will use it in the proof of the central limit theorem. In the remainder of this section, we will reserve the symbol i for the imaginary unit $\sqrt{-1}$.

Definition 9.3

Given a real-valued random variable X, define the *characteristic function* of X by

$$\phi_X(t) \triangleq E[e^{iXt}] = \int_\Omega e^{iX(\omega)t}\, dP(\omega)$$

for $t \in \mathbb{R}$.

If X is a continuous random variable with pdf $f(x)$, we can compute its characteristic function by $\int e^{ixt} f(x)\, dx$. If X is a discrete random variable with pmf $p(n)$, the characteristic function is equal to $\sum_n e^{int} p(n)$.

Example 9.2.1 (Characteristic Function of Bernoulli Random Variable)
Suppose X follows a Bernoulli distribution with parameter p, where $P(X = 0) = 1 - p$ and $P(X = 1) = p$. The characteristic function of X is given by

$$\phi_X(t) = (1 - p)e^{i \cdot 0 \cdot t} + pe^{it} = 1 - p + pe^{it}.$$

Example 9.2.2 (Characteristic Function of Uniform Random Variable)
Suppose X is a continuous random variable that is uniformly distributed on the interval $[-a, a]$, where a is a positive constant. The characteristic function of X is

$$\phi_X(t) = \frac{1}{2a} \int_{-a}^{a} e^{ixt}\, dx = \frac{1}{2a} \frac{e^{iat} - e^{-iat}}{it} = \frac{\sin(at)}{at},$$

for $t \neq 0$. When $t = 0$, $\phi_X(0)$ is equal to 1.

Table 9.1 includes the characteristic functions of some commonly encountered random variables. The second column in the table indicates the support of the distribution, which is range of k in the discrete case or the range of x in the continuous case. The third column lists the probability density function or the probability mass function, while the fourth column shows the characteristic function.

Table 9.1 Table of characteristic functions

Distribution	Support	Pdf/Pmf	Characteristic function
Point mass	$\{a\}$	Dirac delta $\delta_a(x)$	e^{ita}
Bernoulli	$\{0, 1\}$	$(1-p)^{1-k}p^k$	$1-p+pe^{it}$
Binomial	$\{0, 1, \ldots, n\}$	$\binom{n}{k}(1-p)^{n-k}p^k$	$(1-p+pe^{it})^n$
Geometric	$\mathbb{N}$	$(1-p)^{k-1}p$	$pe^{it}/(1-(1-p)e^{it})$
Negative binomial	$\{0, 1, 2\ldots\}$	$\binom{k+r-1}{k}(1-p)^k p^r$	$p^r/(1-(1-p)e^{it})^r$
Poisson	$\{0, 1, 2, \ldots\}$	$\lambda^k e^{-\lambda}/k!$	$\exp(\lambda(e^{it}-1))$
Uniform	$[0, a]$	$1/a$	$(e^{iat}-1)/(ita)$
Gaussian	$\mathbb{R}$	$e^{-x^2/(2\sigma^2)}/\sqrt{2\pi\sigma^2}$	$e^{-\sigma^2 t^2/2}$
Gamma	$\{x : x \geq 0\}$	$x^{\alpha-1}e^{-x/\beta}/(\Gamma(\alpha)\beta^\alpha)$	$(1-i\beta t)^{-\alpha}$
Exponential	$\{x : x \geq 0\}$	$\lambda e^{-\lambda x}$	$\lambda/(\lambda-it)$
Chi-square	$\{x : x \geq 0\}$	$x^{k/2-1}e^{-x/2}/(\Gamma(k/2)2^{k/2})$	$(1-2it)^{-k/2}$
Laplace	$\mathbb{R}$	$\frac{1}{2b}\exp(-|x|/b)$	$1/(1+b^2t^2)$
Cauchy	$\mathbb{R}$	$b/(\pi(1+b^2x^2))$	$e^{-|t|/b}$

9.2.1 Properties of Characteristic Functions

We prove the basic properties of characteristic function in the next theorem.

Theorem 9.3

The characteristic function of a real-valued random variable X satisfies the following properties

1. *(Boundedness)* $|\phi_X(t)| \leq 1$ and $\phi_X(0) = 1$.
2. *(Conjugate symmetry)* $\phi_X(-t) = \phi_X^*(t)$, where $\phi_X^*(t)$ stands for the complex conjugate of $\phi_X(t)$.
3. *If X and Y are independent random variables, then*

$$\phi_{X+Y}(t) = \phi_X(t)\phi_Y(t).$$

4. $\phi_X(t)$ *is uniformly continuous.*

Proof

(1) The property $\phi_X(0) = 1$ follows from the fact that $E[e^0] = E[1] = 1$. To show that $|\phi_X(t)| \leq 1$, we can apply the triangle inequality,

$$\left| \int e^{iX(\omega)t}\, dP(\omega) \right| \leq \int |e^{iXt}|\, dP = \int 1\, dP = 1.$$

(2) For any $t \in \mathbb{R}$, we have $\phi_X(-t) = E[e^{-iXt}] = E[e^{iXt}]^* = \phi_X^*(t)$. This follows from the fact that integral of the complex conjugate of a function is equal to the complex conjugate of the integral.

(3) Since e^{iXt} and e^{iYt} are independent complex random variables, we have

$$E[e^{i(X+Y)t}] = E[e^{iXt}e^{iYt}] = E[e^{iXt}]E[e^{iYt}].$$

Therefore, we obtain the formula $\phi_{X+Y}(t) = \phi_X(t)\phi_Y(t)$.

(4) Applying the triangle inequality yields the upper bound

$$|\phi_X(t+h) - \phi_X(t)| \leq \int |e^{iX(t+h)} - e^{iXt}|\, dP = \int |e^{iXh} - 1|\, dP.$$

Since the integrand is bounded by 2, we can apply the dominated convergence theorem (Theorem 7.7) to obtain

$$\lim_{h \to 0} |\phi_X(t+h) - \phi_X(t)| \leq \lim_{h \to 0} \int |e^{iXh} - 1|\, dP = \int \lim_{h \to 0} |e^{iXh} - 1|\, dP = 0.$$

Because the above limit does not depend on the value of t, we can find a sufficiently small h such that

$$|\phi_X(t+h) - \phi_X(t)| \leq \delta$$

for all $t \in \mathbb{R}$ given any $\delta > 0$. This proves that $\phi_X(t)$ is uniformly continuous. ∎

The last property implies that a small change in the argument of the characteristic function results in a small change in the value of the function. This ensures that the characteristic function does not exhibit large fluctuations.

9.2.2 Inversion Formula

The characteristic function contains complete information about the probability distribution of a random variable. Indeed, we can recover the cumulative distribution function of a random variable from its characteristic function. We first derive a useful estimate for the absolute value of $e^{ix} - 1$.

Theorem 9.4
$|e^{ix} - 1| \leq |x|$ for all $x \in \mathbb{R}$.

Proof We can obtain the estimate $|e^{ix} - 1| \le |x|$ by considering the integral of the complex exponential function on the real number line. For fixed $x > 0$, we have

$$\int_0^x e^{iu}\,du = \left[\frac{e^{iu}}{i}\right]_0^x = \frac{e^{ix} - 1}{i}.$$

Hence, by the triangle inequality, we obtain $|e^{ix} - 1| \le \int_0^x |e^{iu}|\,du = x$.

For negative x, we can write

$$|e^{ix} - 1| = |e^{ix}||1 - e^{i(-x)}| = |e^{i(-x)} - 1| \le -x.$$

Therefore, we have $|e^{ix} - 1| \le |x|$ for any $x \in \mathbb{R}$.   ∎

Theorem 9.5 (Inversion Formula)
Let X be a random variable on a probability space $(\Omega, \mathscr{F}, P)$, and $\phi_X(t)$ be the characteristic function of X. Denote the push-forward measure of P by μ. For $a < b$, we have

$$\mu((a, b)) + \frac{\mu(\{a\})}{2} + \frac{\mu(\{b\})}{2} = \frac{1}{2\pi} \lim_{T \to \infty} \int_{-T}^T \frac{e^{-ita} - e^{-itb}}{it} \phi_X(t)\,dt.$$

On the left-hand side of the equation in Theorem 9.5, (a, b) denotes an open interval, and $\mu(\{a\})$ and $\mu(\{b\})$ are probabilities at the points a and b, respectively. This formulation is necessary to account for any possible discontinuities in the cumulative distribution function. If the cdf is continuous, then both $\mu(\{a\})$ and $\mu(\{b\})$ are zero, and $\mu((a, b)) = \mu([a, b])$. The limit on the right-hand side is the Cauchy principal value of the integral, and its existence is part of the theorem.

In the following proof, we will need to use the Dirichlet integral

$$\lim_{T \to \infty} \int_0^T \frac{\sin u}{u}\,du = \frac{\pi}{2}.$$

This limit is obtained from the Riemann integral of the function $(\sin u)/u$.

Proof We begin by considering the integral for fixed T:

$$I_T \triangleq \int_{-T}^T \frac{e^{-ita} - e^{-itb}}{it} \phi_X(t)\,dt$$

$$= \int_{-T}^T \int_{\mathbb{R}} \frac{e^{-ita} - e^{-itb}}{it} e^{itx}\,d\mu(x)dt.$$

Using the estimate from Theorem 9.4, we can bound the integrand as follows:

$$\left| \frac{e^{-ita} - e^{-itb}}{it} e^{itx} \right| = \left| \frac{1}{t}(e^{it(b-a)} - 1) \right| \le \frac{1}{|t|}|t|(b-a) = b - a.$$

Therefore, we can apply Fubini theorem (Theorem 8.5) to exchange the order of integrations,

$$I_T = \int_{\mathbb{R}} \int_{-T}^{T} \frac{e^{-ita} - e^{-itb}}{it} e^{itx} \, dt \, d\mu(x). \tag{9.1}$$

The integrand of the inner in (9.1) is given by

$$\frac{1}{it}\left(e^{-it(a-x)} - e^{-it(b-x)}\right)$$

$$= \frac{-i}{t}\Big(\cos(t(a-x)) - i\sin(t(a-x)) - \cos(t(b-x)) + i\sin(t(b-x))\Big)$$

$$= \frac{1}{t}\Big(-i\cos(t(a-x)) - \sin(t(a-x)) + i\cos(t(b-x)) + \sin(t(b-x))\Big).$$

We observe that the imaginary part is an odd function of t, while the real part is an even function. Since we are integrating over the symmetric interval $[-T, T]$, the integral of the imaginary part vanishes. Let $f(x, T)$ denote the inner integral in (9.1). Then we have

$$f(x, T) = \int_{-T}^{T} \frac{1}{t}(\sin(t(b-x)) - \sin(t(a-x))) \, dt$$

$$= 2 \int_{0}^{T} \frac{1}{t}\sin(t(b-x)) \, dt - 2 \int_{0}^{T} \frac{1}{t}\sin(t(a-x)) \, dt.$$

Using a change of variable, we simplify this to

$$f(x, T) = 2 \int_{0}^{(b-x)T} \frac{\sin(u)}{u} \, du - 2 \int_{0}^{(a-x)T} \frac{\sin(u)}{u} \, du = 2 \int_{(a-x)T}^{(b-x)T} \frac{\sin(u)}{u} \, du.$$

We consider four different cases.

Case 1: $a < x < b$. We have $(b - x)T \to \infty$ and $(a - x)T \to -\infty$ as $T \to \infty$. Hence $f(x, T) \to 2\pi$ as $T \to \infty$.

Case 2: $x = a$. The upper boundary $(b-x)T$ tends to infinity while $(a-x)T = 0$. In this case, $f(x, T) \to \pi$ as $T \to \infty$.

Case 3: $x = b$. The lower boundary $(a-x)T$ tends to $-\infty$, and $(b-x)T$ equals 0. We have $f(x, T) \to \pi$ as $T \to \infty$.

Case 4:   When $x < a$ or $x > b$, then either the upper and lower boundaries $(b - x)T$ and $(a - x)T$ approach ∞ together, or they approach $-\infty$ together. The integral has limit 0.

Therefore, we obtain

$$\lim_{T \to \infty} f(x, T) = \begin{cases} 2\pi & \text{if } x \in (a, b), \\ \pi & \text{if } x = a \text{ or } x = b, \\ 0 & \text{otherwise.} \end{cases}$$

Finally, we note that the integral $\int_0^v \sin(u)/u \, du$ converges as $v \to \infty$, and the integral as a function of v is a continuous function. This implies that $|f(x, T)|$ is bounded by a constant for all x and for all T. Therefore, we can apply the complex version of the dominated convergence theorem (Theorem 7.6) to justify the following steps:

$$\lim_{T \to \infty} I_T = 2\pi \int \mathbf{1}_{(a,b)}(x) \, d\mu(x) + \pi \int \mathbf{1}_{\{a\}}(x) \, d\mu(x) + \pi \int \mathbf{1}_{\{b\}}(x) \, d\mu(x)$$

$$= 2\pi\mu((a, b)) + \pi\mu(\{a\}) + \pi\mu(\{b\}).$$

Dividing both sides by 2π, we obtain the inversion formula in the theorem.   ∎

Using the inversion formula, we can show that the characteristic function of a random variable uniquely determines the cumulative distribution function.

> **Theorem 9.6 (Uniqueness Theorem)**
> *Let X and Y be random variables with characteristic function $\phi_X(t)$ and $\phi_Y(t)$, respectively. If $\phi_X(t) = \phi_Y(t)$ for all $t \in \mathbb{R}$, then X and Y have the same distribution.*

Proof Let $F_X(x)$ and $F_Y(y)$ denote the cumulative distribution functions of X and Y, respectively. If F_X and F_Y are both continuous at a and b, the inversion formula can be simplified to

$$P(a < X < b) = \frac{1}{2\pi} \lim_{T \to \infty} \int_{-T}^{T} \frac{e^{-iat} - e^{-ibt}}{it} \phi_X(t) \, dt,$$

$$P(a < Y < b) = \frac{1}{2\pi} \lim_{T \to \infty} \int_{-T}^{T} \frac{e^{-iat} - e^{-ibt}}{it} \phi_Y(t) \, dt.$$

Since $\phi_X(t) = \phi_Y(t)$ for all t, we must have $P(a < X < b) = P(a < Y < b)$.

Because there are at most countably many discontinuity points in the cumulative distribution function of a random variable (see Exercise 3.5), the Lebesgue–Stieltjes measures induced by F_X and F_Y agree on a collection of open intervals (a, b), where a and b range over a dense set in $\mathbb{R}$. This collection of intervals generates the Borel algebra on $\mathbb{R}$. Therefore, the two Lebesgue–Stieltjes measures are the same. Consequently, the cdf's of X and Y are the same (see Theorem 3.9). ∎

9.2.3 Computing Moments from Characteristic Function

We can determine the moments of a random variable by differentiating its characteristic function. The proof follows a similar idea as in the case of moment generating function, and it involves justifying the exchange of the order of differentiation and expectation.

> **Theorem 9.7 (Derivatives of Characteristic Function)**
> *Suppose $E[|X|^n] < \infty$. Then the n-th derivative of $\phi_X(t)$ exists and*
> $$\phi_X^{(n)}(t) = E[(iX)^n e^{iXt}].$$
> *In particular, we have $E[X^n] = \frac{1}{i^n}\phi_X^{(n)}(0)$.*

Sketch of Proof We will prove the theorem for $n = 1$ and $n = 2$ only.
Suppose $E[|X|] < \infty$. We want to evaluate the limit

$$\lim_{h \to 0} \frac{\phi_X(t+h) - \phi_X(t)}{h} = \lim_{h \to 0} \int \frac{1}{h}(e^{iX(\omega)(t+h)} - e^{iX(\omega)t})\, dP(\omega).$$

To exchange the order of limit and integral, we show that the integrand is bounded by an integrable function. By Theorem 9.4, we have

$$\frac{1}{|h|}|e^{iX(\omega)(t+h)} - e^{iX(\omega)t}| \le \frac{1}{|h|}|e^{iX(\omega)h} - 1| \le \frac{1}{|h|}|X(\omega)| \cdot |h| = |X(\omega)|.$$

Hence, by the limit version of the dominated convergence theorem (Theorem 7.7), we obtain

$$\phi_X'(t) = \lim_{h \to 0} \frac{\phi_X(t+h) - \phi_X(t)}{h} = \int \lim_{h \to 0} \frac{1}{h}(e^{iX(\omega)(t+h)} - e^{iX(\omega)t})\, dP(\omega)$$

$$= \int iX(\omega)e^{iX(\omega)t}\, dP(\omega)$$

$$= E[iXe^{iXt}].$$

To prove the case $n = 2$, we begin by computing the second derivative of the characteristic function,

$$\lim_{h \to 0} \frac{\phi'(t+h) - \phi'(t)}{h} = \lim_{h \to 0} \int \frac{i}{h} (X(\omega) e^{iX(\omega)(t+h)} - X(\omega) e^{iX(\omega)t}) \, dP(\omega).$$

For each ω, we can bound the absolute value of the integrand by

$$\left| \frac{iX(\omega) e^{iX(\omega)t} (e^{iX(\omega)h} - 1)}{h} \right| = \left| \frac{X(\omega)(e^{iX(\omega)h} - 1)}{h} \right| \leq |X(\omega)|^2.$$

If $E[|X|^2]$ is finite, we can apply dominated convergence theorem to exchange the order of integration and limit,

$$\lim_{h \to 0} \frac{\phi'(t+h) - \phi'(t)}{h} = \int \lim_{h \to 0} \frac{i}{h} (X(\omega) e^{iX(\omega)(t+h)} - X(\omega) e^{iX(\omega)t}) \, dP(\omega)$$

$$= E[(iX)^2 e^{iXt}].$$

This proves the formula for $n = 2$.   ∎

Problems

9.1. Find the moment generating function of the exponential distribution with parameter λ, and use it to determine the k-th moment, for $k \geq 1$.

9.2. Use characteristic function to show that the sum of two independent Poisson random variables is Poisson distributed.

9.3. By using characteristic function, show that the sum of two independent Gamma random variables with distributions $\Gamma(\alpha_1, \beta)$ and $\Gamma(\alpha_2, \beta)$ has distribution $\Gamma(\alpha_1 + \alpha_2, \beta)$.

9.4. Show that the characteristic function of $Y = aX + b$ is $\phi_Y(t) = \phi_X(at) e^{ibt}$.

9.5. Denote the characteristic function of the standard Gaussian distribution by $\psi(t)$.

(a) Show that the imaginary part of $\psi(t)$ is zero, and

$$\psi(t) = \frac{1}{\sqrt{2\pi}} \int_{-\infty}^{\infty} \cos(tx) e^{-x^2/2} \, dx.$$

(b) By applying Theorem 9.7, differentiate $\psi(t)$ and derive $\psi'(t) = -t\psi(t)$.

(c) Derive $\psi(t) = e^{-t^2/2}$ by solving the differential equation in part (b).

(d) Show that the moment of $N(0, 1)$ of order $2k$ is $(2k-1)(2k-3)\cdots 3\cdot 1$.

9.6. Let $X_1, \ldots, X_n$ be i.i.d. Cauchy random variables. Show that $(X_1+\cdots+X_n)/n$ is Cauchy with the same distribution as X_1.

9.7. A random variable Z is called *symmetric* if Z and $-Z$ have the same distribution.

(a) Show that the characteristic function of a symmetric random variable is a real-valued function.

(b) Show that the difference of two i.i.d. random variables is symmetric.

9.8. Prove that $\phi_X(2\pi) = 1$ if and only if $P(X \in \mathbb{Z}) = 1$.

9.9. Let $\phi_X(t)$ denote the characteristic function of a random variable X. Given any n distinct real numbers $t_1 < t_2 < \cdots < t_n$ in ascending order, define an $n \times n$ matrix M by letting the (i, j)-entry be $m_{ij} = \phi_X(t_j - t_i)$. Prove that M is Hermitian and positive semi-definite. That is, show that $M^H = M$ and $\mathbf{v}^H M \mathbf{v} \geq 0$ for all complex vector $\mathbf{v} \in \mathbb{C}^n$. (The superscript ^{H} means conjugate transpose.)

9.10. It is known that the characteristic function of a χ^2 distribution with k degrees of freedom is $(1 - 2it)^{-k/2}$. Find the mean and variance.

9.11. Prove that if $\phi(t)$ is the characteristic function of a random variable, the squared magnitude $|\phi(t)|^2$ is also a characteristic function.

Modes of Convergence

10

To estimate an unknown parameter, we can draw independent samples and obtain an estimator from the samples. A fundamental question is whether the estimates will converge to the intended value if we increase the number of samples. There are several ways to define convergence of random variables. A useful notion of convergence is almost sure (a.s.) convergence, in which the random variables are required to converge to a limit with probability 1.

Another important convergence concept is convergence in the r-th mean, where r is a constant larger than or equal to 1. When $r = 1$, we usually call it convergence in the mean, or L^1 convergence. We have seen in the proof of dominated convergence theorem that, if under the conditions in dominated convergence theorem, the sequence of random variables indeed converges in L^1. When $r = 2$, this is commonly called mean square convergence, or convergence in quadratic mean. It has numerous applications in estimation and filtering.

Convergence in probability is a weaker notion of convergence compared to a.s. convergence and convergence in the mean. This is the mode of convergence in the weak law of large numbers. The most relaxed mode of convergence is convergence in distribution, which is the basis of the central limit theorem. The total variation distance also defines a convergence concept that is useful in large deviation theory. The logical dependency among the aforementioned modes of convergence is depicted below.

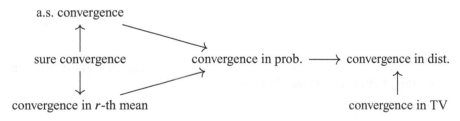

K. Shum, *Measure-Theoretic Probability*, Compact Textbooks in Mathematics, https://doi.org/10.1007/978-3-031-49830-5_10

At the end of this chapter we derive the continuous mapping theorem for a.s. convergence and convergence in probability.

10.1 Convergence Almost Surely and Convergence in Probability

In this section, we discuss the first two fundamental notions of convergence: almost sure convergence and convergence in probability.

Definition 10.1

A sequence of random variables $(X_n)_{n \geq 1}$ defined on a probability space $(\Omega, \mathcal{F}, P)$ is said to converge to X *surely* if $\lim_{n \to \infty} X_n(\omega) = X(\omega)$ for all $\omega \in \Omega$.

A sequence of random variables $(X_n)_{n \geq 1}$ is said to converge to X *almost surely* if there exists an event E with $P(E) = 1$ such that $\lim_{n \to \infty} X_n(\omega) = X(\omega)$ for ω in E. In this case, we write $X_n \xrightarrow{a.s} X$ or $X_n \to X$ with probability 1.

Definition 10.2

A sequence of random variables $(X_n)_{n \geq 1}$ is said to converge to X *in probability* if for any $\epsilon > 0$,

$$P(|X_n - X| > \epsilon) \to 0 \quad \text{as } n \to \infty.$$

In this case, we write $X_n \xrightarrow{P} X$ or $X_n \to X$ in probability.

Almost sure convergence is concerned with the pointwise convergence of random variables. Convergence in probability is about the convergence of sequence of probabilities. The next example illustrates the difference between a.s. convergence and convergence in probability.

Example 10.1.1 (An Example of Convergence in Probability But Not a.s.)
Let X_n be a sequence of independent binary random variables, with probability distribution specified by $P(X_n = 1) = 1/n$ and $P(X_n = 0) = 1 - 1/n$. This sequence converges to 0 in probability. However, by Borel–Cantelli lemma (Theorem 5.9), X_n is equal to 1 infinitely often with probability 1. Hence, with probability 1, it does not converge to 0.

We now establish a result that relates almost sure convergence to convergence in probability. To this end, we use the following characterization of a.s. convergence.

Theorem 10.1
$$X_n \xrightarrow{a.s.} X \iff \forall \epsilon > 0, \ P(|X_n - X| > \epsilon \ i.o.) = 0.$$

Proof ($\Rightarrow$) Suppose $X_n \xrightarrow{a.s.} X$, and let E denote the event that $X_n(\omega) \to X(\omega)$. By assumption, we have $P(E) = 1$.

Fix $\epsilon > 0$. If the sequence $(X_n(\omega))_{n \geq 1}$ converges to $X(\omega)$, there exists a sufficiently large integer N such that $|X_n(\omega) - X(\omega)| \leq \epsilon$ for all $n \geq N$. Hence, if ω is an outcome in Ω such that $|X_n(\omega) - X(\omega)| > \epsilon$ for infinitely many n, we must have $\omega \in E^c$. Since $P(E^c) = 0$, we obtain $P(|X_n - X| > \epsilon \ i.o.) = 0$ as desired.

($\Leftarrow$) Suppose $P(|X_n - X| > \epsilon \ i.o.) = 0$ for all $\epsilon > 0$. By the definition of convergence of sequence, we have

$$X_n(\omega) \to X(\omega) \iff \forall k \in \mathbb{N}, \ |X_n(\omega) - X(\omega)| \leq \frac{1}{k} \text{ for all sufficient large } n.$$

Hence, the event that the sequence $(X_n(\omega))_{n \geq 1}$ fails to converge to $X(\omega)$ is contained in the countable union

$$\bigcup_{k=1}^{\infty} \{\omega : |X_n(\omega) - X(\omega)| > \frac{1}{k} \ i.o.\}.$$

Since all the events on the right are null sets, we have $P(X_n \nrightarrow X) = 0$. $\blacksquare$

The event $\{\omega : |X_n(\omega) - X(\omega)| > \epsilon \ i.o.\}$ in the last theorem can be written in terms of limsup,

$$\{\omega : |X_n(\omega) - X(\omega)| > \epsilon \ i.o.\} = \bigcap_{n=1}^{\infty} \bigcup_{m \geq n} \{\omega : |X_m(\omega) - X(\omega)| > \epsilon\}.$$

By using the semi-continuity property of probability measure, we can alternately express Theorem 10.1 as

$$X_n \xrightarrow{a.s.} X \iff \forall \epsilon > 0, \ \lim_{n \to \infty} P\left(\bigcup_{m \geq n} \{\omega : |X_m(\omega) - X(\omega)| > \epsilon\}\right) = 0.$$

$$(10.1)$$

This equation is useful to check almost sure convergence, as it allows us to focus on the behavior of the tails of the sequence.

Using (10.1), one can readily derive the next theorem.

Theorem 10.2

$X_n \xrightarrow{a.s} X$ *implies* $X_n \xrightarrow{P} X$.

Proof Assume $(X_n)_{n \geq 1}$ converges to X almost surely. By (10.1), we obtain, for all $\epsilon > 0$,

$$0 = \lim_{n \to \infty} P(\cup_{m=n}^{\infty} \{|X_m - X| > \epsilon\}) \geq \lim_{n \to \infty} P(\{|X_n - X| > \epsilon\}).$$

We have inequality in the last step because the event becomes a smaller set. Thus,

$$\lim_{n \to \infty} P(\{|X_n - X| > \epsilon\}) = 0.$$

Because it holds for any $\epsilon > 0$, we have convergence in probability.   ∎

We need the Markov inequality to prove the next theorem.

Theorem 10.3   (Markov Inequality)
If X is a nonnegative random variable with finite expectation, then for any $\epsilon > 0$, we have

$$P(X \geq \epsilon) \leq \frac{E[X]}{\epsilon}.$$

Proof The indicator function $\mathbf{1}_{[\epsilon,\infty)}(x)$ is less than or equal to the linear function $g(x) = x/\epsilon$ for all $x \geq 0$. By the monotonic property of integral, we obtain

$$P(X \geq \epsilon) = E[\mathbf{1}_{[\epsilon,\infty)}(X)]$$

$$\leq E[X/\epsilon].$$

By linearity of expectation, we obtain $E[X/\epsilon] = E[X]/\epsilon$, and thus proving the Markov inequality.   ∎

The next theorem shows that if the means of a sequence of nonnegative random variables converge to 0 sufficiently fast, then the sequence must converge to 0 almost surely. The proof illustrates how to use Theorem 10.1 together with the Borel–Cantelli lemma to establish almost sure convergence.

Theorem 10.4
Suppose $(X_n)_{n=1}^{\infty}$ is a sequence of nonnegative random variable. If $\sum_{n=1}^{\infty} E[X_n] < \infty$, then the sequence $(X_n)_{n=1}^{\infty}$ converges to 0 almost surely.

Proof Fix a positive real number ϵ. By Markov inequality, we have

$$\sum_{n=1}^{\infty} P(X_n > \epsilon) \leq \sum_{n=1}^{\infty} \frac{E[X_n]}{\epsilon}.$$

Since $\sum_{n=1}^{\infty} E[X_n]$ is finite, the summation on the left-hand side is also finite. By the Borel–Cantelli lemma (Theorem 5.8), we have $P(X_n > \epsilon \ i.o.) = 0$. Since this holds for all $\epsilon > 0$, we can apply Theorem 10.1 and conclude that $X_n \xrightarrow{a.s.} 0$. Note that we need not assume independence in this proof. ∎

10.2 Convergence in the Mean

Definition 10.3

For $r \geq 1$, $(X_n)_{n\geq 1}$ is said to converge to X *in the r-th mean* (or *in the L^r norm*) if

$$E[\, |X_n - X|^r \,] \to 0 \quad \text{as } n \to \infty.$$

When $r = 1$, we say that X_n converges to X *in the mean*. When $r = 2$, we say that X_n converges to X *in mean square* or *in quadratic mean*. Other notation for mean square convergence include $X_n \xrightarrow{m.s.} X$ and l.i.m. $X_n = X$.

From Theorem 7.5, we see that whenever the dominated convergence theorem can be applied, the sequence of random variables converges in the mean with $r = 1$.

Convergence in the mean and convergence almost surely do not have a direct relationship with each other. As the next two examples show, convergence in the mean does not imply almost sure convergence and *vice versa*.

Example 10.2.1 (Almost Sure Convergence But Not in the Mean)
Let $\Omega = [0, 1]$, and P be the Lebesgue measure on $[0, 1]$. For $n = 1, 2, 3, \ldots$, define

$$X_n(\omega) \triangleq n \cdot \mathbf{1}_{[0,1/n]}(\omega).$$

The sequence of random variables $(X_n)_{n\geq 1}$ does not converge to 0 in the mean, because

$$E[X_n - 0] = \int_{[0,1/n]} n \, dP = 1$$

for all n. However, the sequence converge to 0 a.s., because for each $\omega > 0$, the sequence $(X_n(\omega))_{n\geq 1}$ will eventually be equal to 0.

Example 10.2.2 (Convergence in the Mean But Not Almost Surely)
Consider the probability space as in the previous example. Define a sequence of events A_k, for $k \geq 1$,

$$A_1 \triangleq \Omega$$

$$A_2 \triangleq [0, 1/2], \; A_3 \triangleq [1/2, 1],$$

$$A_4 \triangleq [0, 1/4], \; A_5 \triangleq [1/4, 1/2], \; A_6 \triangleq [1/2, 3/4], \; A_7 \triangleq [3/4, 1], \ldots$$

For $2^k \leq i \leq 2^{k+1} - 1$, the union of A_i's equals the whole sample space $[0, 1]$, and $P(A_i) = 1/2^k$. For $k \geq 1$, let X_k be the indicator function $\mathbf{1}_{A_k}$. The sequence $(X_k)_{k\geq 1}$ does not converge to 0 a.s., because for each ω, $X_k(\omega) = 1$ for infinitely many k. On the other hand, X_k's converge to 0 in the mean.

However, we can show that convergence in the mean is stronger than convergence in probability.

Theorem 10.5

Convergence in L^r norm implies convergence in probability.

Proof Suppose $(X_n)_{n\geq 1}$ converges to X in the r-th mean for some positive number r, i.e., $E[|X_n - X|^r] \to 0$ as $n \to \infty$. For any $\epsilon > 0$, we apply the Markov inequality to obtain

$$P(|X_n - X| > \epsilon) = P(|X_n - X|^r > \epsilon^r) \leq \frac{E[|X_n - X|^r]}{\epsilon^r} \to 0$$

as $n \to \infty$. This proves that the sequence $(X_n)_{n=1}^{\infty}$ is converging to X in probability. ∎

10.3   Convergence in Distribution and in Total Variation

Definition 10.4 (Convergence in Distribution)

Let $(F_n)_{n=1}^{\infty}$ be a sequence of cumulative distribution functions, and F be another cumulative distribution function. We say that $(F_n)_{n=1}^{\infty}$ *converges in distribution*, or *in law*, if $\lim_{n\to\infty} F_n(x) = F(x)$ at every continuity point of $F(x)$. We use the notation

$$F_n \xrightarrow{D} F \text{ or } F_n \xrightarrow{L} F$$

for convergence in distribution.

Given a sequence of probability measures μ_n's and another probability measure μ, all defined on the real number line, we say that $(\mu_n)_{n=1}^{\infty}$ *converges to μ in distribution* if the corresponding Stieltjes measure functions $F_n(x) = \mu_n((-\infty, x])$ converge to $F(x) = \mu((-\infty, x])$ in distribution.

A sequence of random variables X_n's, for $n = 1, 2, 3, \ldots$ is said to be converging to a random variable X *in distribution*, or *in law*, if the cumulative distribution functions of X_n's converge to the cumulative distribution function of X in distribution.

Definition 10.5 (Convergence in Total Variation)

A sequence of probability measures $(P_n)_{n=1}^{\infty}$ defined on a common measurable space is said to *converge in total variation* to a probability measure P if $d_{TV}(P_n, P) \to 0$ as $n \to \infty$.

A sequence of random variables $(X_n)_{n\geq 1}$ is said to be convergent in total variation if the corresponding image measures on the measure space $(\mathbb{R}, \mathscr{B}(\mathbb{R}))$ converges in total variation.

There is an important difference between these two modes of convergence. In convergence in total variation distance, the probability measures must be defined on the same probability space, while in convergence in distribution, the probability measures may be defined on different probability spaces. In general, convergence in total variation implies convergence in distribution.

Theorem 10.6

If a sequence of probability distributions converges in total variation, then it converges in distribution.

We defer the proof to Chap. 14 after the introduction to weak convergence (See Theorem 14.4).

In the definition of convergence in distribution, we do not need to consider the discontinuity points of the target cdf $F(x)$. To see why, consider a sequence of positive real numbers $x_1, x_2, x_3 \ldots$ converging to 0 from the right. Let $F_n(x) = \mathbf{1}_{[x_n, \infty)}(x)$ be the cumulative distribution function of the Dirac measure that concentrates at the point x_n, for $n \geq 1$. We naturally regard this sequence of probability distributions as converging to the Dirac measure that concentrates at $x = 0$. We can check that $F_n(0) = 0$ for all n, hence $\lim_{n\to\infty} F_n(x) = \mathbf{1}_{(0,\infty)}(x)$. The limit is not a Stieltjes measure function as it is not continuous from the right at $x = 0$. However, the target cdf is $F(x) = \mathbf{1}_{[0,\infty)}(x)$, whose value at $x = 0$ is different from the limit $\lim_{n\to\infty} F_n(x)$. Because the discontinuity point $x = 0$ is neglected in the definition of convergence in distribution, we do have $F_n(x) \to F(x)$ in distribution.

Example 10.3.1 (Empirical Distribution)

Suppose we are given n data points $a_1, a_2, \ldots, a_n$, obtained from a sampling process. To investigate the distribution of the data points, we can consider the empirical distribution defined by the cumulative distribution function

$$F_n(x) \triangleq \frac{1}{n} \sum_{k=1}^{n} \mathbf{1}_{[a_k, \infty)}(x).$$

The graph of $F_n(x)$ is a staircase graph in which each step has size $1/n$, and the steps occur at the points $a_1, a_2, \ldots, a_n$.

Suppose that data points are $\{1/n, 2/n, \ldots, n/n\}$, where n is a positive integer. The cdf of the empirical distribution can be written as

$$F_n(x) = \begin{cases} 0 & \text{if } x < 0 \\ \frac{1}{n}\lfloor xn \rfloor & \text{if } 0 \leq x \leq 1 \\ 1 & \text{if } x > 1. \end{cases}$$

For any $x \in [0, 1]$, we can check that $\lim_{n \to \infty} F_n(x) = \lim_{n \to \infty} \frac{1}{n}\lfloor xn \rfloor = x$. Therefore,

$$\lim_{n \to \infty} F_n(x) = \begin{cases} 0 & \text{if } x < 0 \\ x & \text{if } 0 \leq x \leq 1 \\ 1 & \text{if } x > 1, \end{cases}$$

which is the cdf of the uniform distribution between 0 and 1.

However, we do not have convergence in total variation in this example. Let μ_n be the Lebesgue–Stieltjes measure associated to $F_n(x)$, for $n \geq 1$, and μ be the Lebesgue–Stieltjes measure associated to $F(x)$. If we take A_n to be the discrete set $\{1/n, 2/n, \ldots, n/n\}$, then $|\mu_n(A_n) - \mu(A_n)| = 1$, for all $n \geq 1$.

The last example is an example of convergence in distribution but not convergence in total variation.

The cumulative distribution function is used in the definition of convergence in distribution. However, in practice, we can also check convergence in distribution using probability density function, as demonstrated in the following example.

Example 10.3.2 (Convergence of Pdf's Implies Convergence in Total Variation)

Suppose $(X_n)_{n \geq 1}$ is a sequence of continuous random variables with corresponding probability density functions $f_n(x)$. Suppose $f_n(x)$'s converge to a probability density function $f(x)$, for almost all x relative to the Lebesgue measure on $\mathbb{R}$. By the Scheffé lemma (Exercise 7.6), we have $\int |f_n - f| \to 0$ as $n \to \infty$, which implies convergence in total variation. Since convergence in total variation implies convergence in distribution in general, we can conclude that the sequence of random variables X_n converges in distribution.

In particular, suppose $\mu_n \to \mu$ and $\sigma_n^2 \to \sigma^2$. Then the probability density functions of the Gaussian distribution with mean μ_n and variance σ_n^2 converge pointwise to the probability density functions of the Gaussian distribution with mean μ and variance σ^2, as $n \to \infty$. By the arguments in the previous paragraph, we conclude that $N(\mu_n, \sigma_n^2) \xrightarrow{D} N(\mu, \sigma^2)$.

The relationship between convergence in probability and convergence in distribution is given by the following example and theorem, which together say that convergence in probability is strictly stronger than convergence in distribution.

Example 10.3.3 (Convergence in Distribution But Not in Probability)
Consider two independent random variables X and Y. Each of them is distributed according to Ber($1/2$). Define X_n to be equal to X for all $n \geq 1$. We trivially have X_n converging to Y in distribution, because all distributions are identical. However, X_n does not converge to Y in probability, because $P(|X_n - Y| = 1) = 1/2$ for all n. Hence, $P(|X_n - Y| \geq 0.99) = 1/2$ for all n, and the sequence does not converge to Y in probability.

Theorem 10.7
Suppose X_n, for $n \geq 1$, is a sequence of random variables converging to X in probability, then X_n converges to X in distribution.

Proof Let $F_n(x)$ be the cumulative distribution function of X_n, for $n \geq 1$, and let $F(x)$ be the cumulative distribution function of X. Suppose that for each $\delta > 0$, we have $\lim_{n \to \infty} P(|X_n - X| \geq \delta) = 0$. We want to prove

$$\lim_{n \to \infty} P(X_n \leq a) = P(X \leq a),$$

for each a that such that $F(x)$ is continuous at $x = a$.

Suppose $\delta > 0$ and a constant a is given. The key of the proof is the following inequalities:

$$P(X_n \leq a) \leq P(X \leq a + \delta) + P(|X_n - X| > \delta) \tag{10.2}$$

$$P(X \leq a - \delta) \leq P(X_n \leq a) + P(|X_n - X| > \delta). \tag{10.3}$$

To see the first inequality, note that if $X_n \leq a$ and the difference between X_n and X is less than or equal to δ, then we must have $X \leq a + \delta$. This yields (10.2). The second inequality (10.3) can be derived similarly.

Take n approaching infinity and use the assumption that X_n converges to X in probability, we have

$$\limsup_{n \to \infty} P(X_n \leq a) \leq P(X \leq a + \delta)$$

and

$$P(X \leq a - \delta) \leq \liminf_{n \to \infty} P(X_n \leq a)$$

for each $\delta > 0$. By taking $\delta \to 0$, we get

$$P(X < a) \leq \liminf_{n \to \infty} P(X_n \leq a) \leq \limsup_{n \to \infty} P(X_n \leq a) \leq P(X \leq a).$$

If $F(x)$ is continuous at $x = a$, then all of the inequalities above are equality, and hence $\lim_{n \to \infty} P(X_n \le a) = P(X \le a)$.   ∎

There is a partial converse to Theorem 10.7: if X_n converges in distribution to a constant c, then X_n converges in distribution to the same constant.

We conclude this section with Skorokhod's representation theorem, which provides a relationship between convergence in distribution and almost sure convergence.

Theorem 10.8 (Skorokhod's Representation Theorem)

If $X_n \overset{D}{\longrightarrow} X$, then we can construct random variables Y and Y_n, for $n \ge 1$, on a common probability space, such that $Y_n \overset{a.s.}{\longrightarrow} Y$, Y_n has the same distribution as X_n for all n, and Y has the same distribution as X.

Proof We need to construct random variables Y_n's and Y, defined on a common probability space, so that (i) $Y_n(\omega)$'s converge to $Y(\omega)$ almost surely, (ii) Y_n and X_n are identically distributed for all n, and (iii) Y and X are identically distributed. The proof is an application of coupling argument.

To construct Y_n and Y, we take the probability space $([0, 1], \mathscr{B}([0, 1]), \lambda)$, where λ represents the Lebesgue measure on $[0, 1]$, as the common probability space. Using the distribution function $F_n(x)$, we define a function $Y_n : [0, 1] \to \mathbb{R}$, for $n = 1, 2, 3, \ldots$, by

$$Y_n(\omega) \triangleq \sup\{y : F_n(y) < \omega\}, \tag{10.4}$$

and define

$$Y(\omega) \triangleq \sup\{y : F(y) < \omega\}. \tag{10.5}$$

We take the supremums in the definitions of Y_n and Y to handle the potential discontinuities of the distribution functions F_n and F. If F_n and F are continuous and one-to-one such that the inverses F_n^{-1} and F^{-1} exist, then we can simply define Y_n and Y by the inverse functions.

One can verify that $\{\omega : Y_n(\omega) \le y\} = \{\omega : \omega \le F_n(y)\}$ and $\{\omega : Y(\omega) \le y\} = \{\omega : \omega \le F(y)\}$. These calculations show that $F_n(y)$ is the cdf of Y_n, and $F(y)$ is the cdf of Y. Hence, Y_n and X_n are identically distributed for all n, and the random variables X and Y are also identically distributed.

To show that $Y_n(\omega) \to Y(\omega)$ for all $\omega \in (0, 1)$, we further define

$$Y_n^+(\omega) \triangleq \inf\{y : F_n(y) > \omega\} \text{ for } n \ge 1, \text{ and}$$

$$Y^+(\omega) \triangleq \inf\{y : F(y) > \omega\}.$$

By construction, we have $Y_n \leq Y_n^+$ and $Y \leq Y^+$. But Y_n and Y_n^+ are equal almost surely for all n, and Y and Y^+ are also equal almost surely.

We fix ω in $(0, 1)$. Take a continuity point y_0 of $F(y)$ such that $y_0 > Y^+(\omega)$. Then $F(y_0) > \omega$, and for all sufficiently large n, we have $F_n(y_0) > \omega$. From the definition of Y_n^+, we have $Y_n^+(\omega) < y_0$ for all sufficiently large n. Hence, $\limsup Y_n^+(\omega) \leq y_0$. Because F has at most countably many discontinuity points (see Exercise 3.5), we can take a sequence of $(y_i)_{i=1}^{\infty}$ such that F is continuous at y_i for all i, and $y_i \searrow Y^+(\omega)$. This gives

$$\limsup_n Y_n^+(\omega) \leq Y^+(\omega).$$

Similarly, we can prove that $\liminf_n Y_n(\omega) \geq Y(\omega)$. For each ω, we have the inequalities

$$Y(\omega) \leq \liminf_n Y_n(\omega) \leq \limsup_n Y_n^+(\omega) \leq Y^+(\omega).$$

Since Y and Y^+ are equal almost surely, Y_n^+ converges to Y almost surely. Therefore, we have $Y_n(\omega) \to Y(\omega)$ for ω in a subset of $(0, 1)$ with probability 1. ∎

10.4 Convergence of Random Vectors

In this section, we formally establish the relationship between the convergence of random vectors and the convergence of their component random variables.

Consider measurable functions that take value in $\mathbb{R}^d$. Let $(\Omega, \mathscr{F}, P)$ denote a probability space and $\mathbf{V}(\omega)$ denote a measurable mapping from $(\Omega, \mathscr{F})$ to $(\mathbb{R}^d, \mathscr{B}(\mathbb{R}^d))$. We can prove that a vector function is measurable if and only if each component is a measurable function. This result in measure theory is known as the "coordinate-wise convergence theorem."

Theorem 10.9

Let $(\Omega, \mathscr{F})$ be a measurable space and

$$\mathbf{V}(\omega) \triangleq (X_1(\omega), X_2(\omega), \ldots, X_d(\omega))$$

be a mapping from Ω to $\mathbb{R}^d$, where $X_i(\omega)$ is the i-th component function. Then $\mathbf{V}(\omega)$ is $(\mathscr{F}, \mathscr{B}(\mathbb{R}^d))$-measurable if and only if $X_i(\omega)$ is $(\mathscr{F}, \mathscr{B}(\mathbb{R}))$-measurable for $i = 1, 2, \ldots, d$.

Proof

($\Leftarrow$) Suppose X_i is measurable for all i. We want to show that $\mathbf{V}^{-1}(B)$ is $\mathscr{F}$-measurable for any Borel set B in $\mathscr{B}(\mathbb{R}^d)$. Since $\mathscr{B}(\mathbb{R}^d)$ is generated by d-dimensional hyper-cubes (see Theorem 2.8), it is sufficient to show that $\mathbf{V}^{-1}(C)$ is $\mathscr{F}$-measurable for any hyper-cube C of the form $(a_1, b_1) \times (a_2, b_2) \times \cdots \times (a_d, b_d)$, where a_i, b_i are real numbers, possibly $\pm\infty$.

For any hyper-cube C, the inverse image $\mathbf{V}^{-1}(C)$ is equal to

$$\{\omega : X_i(\omega) \in (a_i, b_i) \text{ for } i = 1, 2, \ldots, d\} = \cap_{i=1}^{d} \{\omega : X_i(\omega) \in (a_i, b_i)\}.$$

Since each set in the intersection is $\mathscr{F}$-measurable, the set $\mathbf{V}^{-1}(C)$ is also $\mathscr{F}$-measurable.

($\Rightarrow$) Suppose $\mathbf{V}$ is $\mathscr{F}$-measurable. That is, we suppose that $\mathbf{V}^{-1}(B)$ is $\mathscr{F}$-measurable for all Borel sets $B \in \mathscr{B}(\mathbb{R}^d)$. For each $i = 1, 2, \ldots, d$, let A_i to be the "cylinder"

$$A_i = \underbrace{\mathbb{R} \times \cdots \times \mathbb{R}}_{i-1 \text{ times}} \times \underbrace{(a_i, b_i)}_{i\text{-th component}} \times \mathbb{R} \times \cdots \times \mathbb{R},$$

which is a Borel set. Then $\mathbf{V}^{-1}(A_i)$ is $\mathscr{F}$-measurable, and it is equal to $\mathbf{V}^{-1}(A_i) = X_i^{-1}((a_i, b_i))$. Hence $X_i^{-1}((a_i, b_i))$ is $\mathscr{F}$-measurable for all $a_i, b_i \in \mathbb{R}$. ∎

We measure distance in $\mathbb{R}^d$ by the Euclidean norm and use the following notation

$$\|(x_1, x_2, \ldots, x_d)\| \triangleq \sqrt{x_1^2 + x_2^2 + \cdots + x_d^2}.$$

Definition 10.6

Let $(\Omega, \mathscr{F}, P)$ denote a probability space, and let $\mathbf{V}_n(\omega)$ be a sequence of d-dimensional random vectors defined on Ω. The sequence of random vectors $\mathbf{V}_n(\omega)$ is said to *converge almost surely to* $\mathbf{V}(\omega)$ if there is an event E with $P(E) = 1$ such that

$$\lim_{n \to \infty} \mathbf{V}_n(\omega) = \mathbf{V}(\omega) \text{ for all } \omega \in E.$$

We say that $\mathbf{V}_n(\omega)$ *converges in probability to* $\mathbf{V}(\omega)$ if for all $\epsilon > 0$, the probability

$$P(\{\omega : \|\mathbf{V}_n(\omega) - \mathbf{V}(\omega)\| > \epsilon\})$$

approaches zero as $n \to \infty$.

The next theorem gives the relation between convergence as a random vector and convergence in each individual component.

> **Theorem 10.10**
> Let $(\Omega, \mathcal{F}, P)$ be a probability space. For $n = 1, 2, 3, \ldots$, let $\mathbf{V}_n(\omega) = (X_{n1}(\omega), X_{n2}(\omega), \ldots, X_{nd}(\omega))$ be a d-dimensional random vector, and let $\mathbf{V}(\omega) = (X_1(\omega), X_2(\omega), \ldots, X_d(\omega))$ be another d-dimensional random vector. Then
>
> (a) $\mathbf{V}_n \xrightarrow{a.s.} \mathbf{V}$ if and only if $X_{ni} \xrightarrow{a.s.} X_i$ for all $i = 1, \ldots, d$.
> (b) $\mathbf{V}_n \xrightarrow{P} \mathbf{V}$ if and only if $X_{ni} \xrightarrow{P} X_i$ for all $i = 1, \ldots, d$.

Proof

(a) ($\Leftarrow$) Suppose the component functions converge a.s. That is, for each $i = 1, 2, \ldots, d$, we can find an event E_i with $P(E_i) = 1$ such that

$$X_{ni}(\omega) \xrightarrow{a.s.} X_i(\omega)$$

for $\omega \in E_i$. Then in $\cap_{i=1}^d E_i$, we have the convergence of random vector,

$$\lim_{n \to \infty} \mathbf{V}_n(\omega) = \mathbf{V}(\omega) \quad \text{for } \omega \in \bigcap_{i=1}^d E_i.$$

The proof is finished by noting that $P(\cap_{i=1}^d E_i) = 1$.

($\Rightarrow$) Suppose $\mathbf{V}_n \xrightarrow{a.s.} \mathbf{V}$. There is a set E with probability 1 such that $\mathbf{V}_n(\omega) \to \mathbf{V}(\omega)$ for $\omega \in E$. Each component of $\mathbf{V}_n(\omega)$ converges to the corresponding component of $\mathbf{V}$ for $\omega \in E$.

(b) ($\Leftarrow$) Assume that each component of $\mathbf{V}_n$ converges in probability to the corresponding component of $\mathbf{V}$. For any $\omega \in \Omega$,

$$|X_{ni}(\omega) - X_i(\omega)| \le \epsilon/\sqrt{d} \text{ for all } i \Rightarrow \|\mathbf{V}_n(\omega) - \mathbf{V}(\omega)\| \le \epsilon.$$

We thus have the following set inclusion

$$\{\omega : \|\mathbf{V}_n - \mathbf{V}\| > \epsilon\} \subseteq \bigcup_{i=1}^d \{\omega : |X_{ni} - X_i| > \epsilon/\sqrt{d}\}.$$

Taking probability of both sides, we get $P(\|\mathbf{V}_n - \mathbf{V}\| > \epsilon) \to 0$ as $n \to \infty$, which follows from the union bound (Exercise 2.4).

($\Rightarrow$) Assume that $\mathbf{V}_n$ converges to $\mathbf{V}$ in probability as a sequence of random vectors. Consider the i-th component of $\mathbf{V}_n$. Since

$$|X_{ni}(\omega) - X_i(\omega)| \le \|\mathbf{V}_n(\omega) - \mathbf{V}(\omega)\|,$$

we have

$$\{\omega : |X_{ni}(\omega) - X_i(\omega)| > \epsilon\} \subseteq \{\omega : \|\mathbf{V}_n(\omega) - \mathbf{V}(\omega)\| > \epsilon\}.$$

Hence, for any $\epsilon > 0$,

$$\mathbf{V}_n \xrightarrow{P} \mathbf{V} \Rightarrow P(\|\mathbf{V}_n(\omega) - \mathbf{V}(\omega)\| > \epsilon) \to 0 \quad \text{as } n \to \infty$$

$$\Rightarrow P(|X_{ni}(\omega) - X_i(\omega)| > \epsilon) \to 0 \quad \text{as } n \to \infty$$

which implies that, by the definition of convergence in probability, X_{ni} converges to X_i in probability for all i.   ∎

Using Theorem 10.10, we can check whether a sequence of random vectors converges in probability by considering the convergence of each component.

10.5   Application: Continuous Mapping Theorem

In this section we prove two continuous mapping theorems—one for almost sure convergence and one for convergence in probability. The two theorems say that almost sure convergence and convergence in probability are preserved by continuous functions.

Theorem 10.11
Let $(\Omega, \mathscr{F}, P)$ be a probability space, $\mathbf{V}$ a d-dimensional random vector, and $(\mathbf{V}_n)_{n=1}^{\infty}$ a sequence of d-dimensional random vectors. Suppose $f : \mathbb{R}^d \to \mathbb{R}^m$ is a function that is continuous on a set $S_f \subseteq \mathbb{R}^d$ with $P(\mathbf{V} \in S_f) = 1$. Then,

(a) $\mathbf{V}_n \xrightarrow{a.s.} \mathbf{V}$ implies $f(\mathbf{V}_n) \xrightarrow{a.s.} f(\mathbf{V})$.
(b) $\mathbf{V}_n \xrightarrow{P} \mathbf{V}$ implies $f(\mathbf{V}_n) \xrightarrow{P} f(\mathbf{V})$.

Proof We let f_k be the k-th component of the function f, for $k = 1, 2, \ldots, m$.

(a) By Theorem 10.10, it is sufficient to show that for each component function f_k of f, we have $f_k(\mathbf{V}_n) \xrightarrow{a.s.} f_k(\mathbf{V})$.

From the hypotheses in the theorem, there exists a set B with probability 1 such that $\mathbf{V}_n(\omega) \to \mathbf{V}(\omega)$ for all $\omega \in B$. Let A be the pre-image $\mathbf{V}^{-1}(S_f)$. The set A has probability 1 by assumption. Hence, in $A \cap B$, we have $\lim_{n\to\infty} \mathbf{V}_n(\omega) = \mathbf{V}(\omega)$, and $\lim_{n\to\infty} f_k(\mathbf{V}_n(\omega)) = f_k(\mathbf{V}(\omega))$, by the continuity of f_k. Therefore $f_k(\mathbf{V}_n)$ converges to $f_k(\mathbf{V})$ for all ω in $A \cap B$, which has probability 1.

(b) By Theorem 10.10, it is sufficient to show that for each component function f_k of f, we have $f_k(\mathbf{V}_n)$ converging to $f_k(\mathbf{V})$ in probability. In the following we fix an index k.

Let δ and ϵ be positive real numbers. Consider the set

$$A_{\delta,\epsilon} \triangleq \{\mathbf{v} \in \mathbb{R}^d : \exists \mathbf{w} \in \mathbb{R}^d \ s.t. \ \|\mathbf{w} - \mathbf{v}\| < \delta, \ \|f_k(\mathbf{w}) - f_k(\mathbf{v})\| > \epsilon\}.$$

For each vector $\mathbf{v}$ that is not in $A_{\delta,\epsilon}$, we cannot find any vector $\mathbf{w}$ that satisfies $\|\mathbf{w} - \mathbf{v}\| < \delta$ and $\|f_k(\mathbf{w}) - f_k(\mathbf{v})\| > \epsilon$ simultaneously. In set notation, we can express it as

$$\{\omega : \|f_k(\mathbf{V}_n(\omega)) - f_k(\mathbf{V}(\omega))\| > \epsilon\}$$
$$\subseteq \{\omega : \mathbf{V}(\omega) \in A_{\delta,\epsilon}\} \cup \{\omega : \|\mathbf{V}_n(\omega) - \mathbf{V}(\omega)\| \geq \delta\}.$$

Take $\delta = 1/j$ for positive integer j. By the continuity of f, and hence of the component function f_k, the set $A_{1/j,\epsilon} \cap S_f \searrow \emptyset$ as j tends to ∞. By the upper semi-continuous property of measure, we obtain

$$P(\{\omega : \mathbf{V}(\omega) \in A_{1/j,\epsilon}\}) = P(\{\omega : \mathbf{V}(\omega) \in A_{1/j,\epsilon} \cap S_f\}) \to 0, \text{ as } j \to \infty.$$

For an arbitrarily small $\epsilon_1 > 0$, we can find a sufficiently large M such that

$$P(\{\omega : \mathbf{V}(\omega) \in A_{1/\ell,\epsilon}\}) \leq \epsilon_1/2$$

for all $\ell \geq M$.

Since it is assumed that $\mathbf{V}_n \to \mathbf{V}$ in probability, we can find a sufficiently large N such that $P(\{\|\mathbf{V}_\ell(\omega) - \mathbf{V}(\omega)\| \geq \delta\}) \leq \epsilon_1/2$ for all $\ell \geq N$. This yields

$$P(\{\|f_k(\mathbf{V}_\ell(\omega)) - f_k(\mathbf{V}(\omega))\| > \epsilon\}) \leq \epsilon_1/2 + \epsilon_1/2 = \epsilon_1$$

for all $\ell \geq \max(M, N)$. As ϵ_1 is arbitrary, the probability $P(\|f_k(\mathbf{V}_\ell) - f_k(\mathbf{V}))\| > \epsilon)$ converges to 0 as $\ell \to \infty$. $\blacksquare$

Theorem 10.12

(i) *Suppose* $X_n \xrightarrow{a.s.} X$ *and* $Y_n \xrightarrow{a.s.} Y$. *Then*

(a) $X_n + Y_n \xrightarrow{a.s.} X + Y$.

(continued)

Theorem 10.12 (continued)

    (b) $X_n Y_n \xrightarrow{a.s.} XY$.

    (c) $\alpha X_n \xrightarrow{a.s.} \alpha X$ *for any constant* α.

    (d) $X_n/Y_n \xrightarrow{a.s.} X/Y$ *provided that* Y_n *and* Y *are nonzero with*

      probability 1.

  (ii) *If* $X_n \xrightarrow{P} X$ *and* $Y_n \xrightarrow{P} Y$. *Then* (a) *to* (d) *above hold with* "a.s."
     replaced by "P."

Proof We prove part (a) for convergence in probability only. The proofs of the others are similar. Since X_n and Y_n are both converging to X and Y, respectively, in probability. The random vector (X_n, Y_n) converges to the random vector (X, Y) in probability. We apply Theorem 10.11 with the continuous function $f(x, y) = x + y$. (cf. the proof of Theorem 4.6) ∎

Exercise 10.10 gives a partial extension of this theorem.

Example 10.5.1 (Consistent Estimator)
Given n i.i.d. samples $X_1, X_2, \ldots, X_n$ drawn from a probability distribution with fixed but unknown parameter θ. We estimate θ by a function $\hat{\theta}_n(X_1, \ldots, X_n)$ of the samples. The estimator $\hat{\theta}_n$ is called *strongly consistent* if $\hat{\theta}_n(X_1, \ldots, X_n)$ converges to θ a.s. and is called *weakly consistent* if it converges to θ in probability.

Suppose a consistent estimator (either strongly or weakly) of the variance of the distribution is available. If we want to estimate the standard deviation, a natural solution is to take the square root of the estimated variance. The continuous mapping theorem guarantees that the resulting estimate is consistent.

Problems

10.1. Prove that a sequence of random variables $(X_k)_{k \geq 1}$ converges to X a.s. if and only if for each $\epsilon > 0$,

$$\lim_{n \to \infty} P(|X_k - X| \leq \epsilon \text{ for all } k \geq n) = 1.$$

10.2. For $n \geq 1$, let X_n be independent Bernoulli random variable with distribution $\Pr(X_n = 1) = p_n$ and $\Pr(X_n = 0) = 1 - p_n$, where p_n is a real number between 0 and 1.

(a) Show that $(X_n)_{n \geq 1}$ converges to 0 in probability if and only if $\lim_{n \to \infty} p_n = 0$.
(b) Show that $(X_n)_{n \geq 1}$ converges to 0 a.s. if and only if $\lim_{n \to \infty} p_n$ is finite.

10.3. Prove that if a sequence of random variables $(X_k)_{k\geq 1}$ converges in distribution to X that satisfies $P(X = c) = 1$ for some constant c, then $X_k \to X$ in probability.

10.4. Suppose $(X_n)_{n=1}^{\infty}$ is a sequence of random variables and $(a_n)_{n=1}^{\infty}$ is a sequence of real numbers such that $X_n - a_n \xrightarrow{P} 0$ and $\lim_{n\to\infty} a_n = c$. Deduce that $X_n \xrightarrow{P} c$.

10.5. Suppose $X_n \to 0$ in probability. Show that if $|X_n| \leq Y$ for an integrable random variable Y for all n, then $X_n \to 0$ in the mean.

10.6. Suppose the sample space Ω is a countable set. Prove that $\mu_n \xrightarrow{D} \mu$ if and only if $\mu_n(\{x\}) \to \mu(\{x\})$ as $n \to \infty$ for all $x \in \Omega$.

10.7. Let X be a discrete random variable with distribution $P(X = -1) = P(X = 1) = 1/2$, and $(U_n)_{n\geq 1}$ be a sequence of i.i.d. random variable uniformly distributed between 0 and 1. Assume that X and $(U_n)_{n\geq 1}$ are independent. For $n = 1, 2, \ldots$, define

$$Y_n = \begin{cases} X & \text{if } U_n \geq 1/n^2 \\ \pi n & \text{if } U_n < 1/n^2. \end{cases}$$

Determine whether $(Y_n)_{n\geq 1}$ converges to X (i) in probability, (ii) almost surely, (iii) in mean square?

10.8. For each $n = 1, 2, 3, \ldots$, let X_n denote a discrete random variable whose value is uniformly distributed on $\{k^2/n^2 : k = 1, 2, \ldots, n\}$. Prove that the random variables X_n's converge in distribution, and find the distribution in the limit.

10.9. Let X_n be geometrically distributed random variable with pmf $P(X_n = k) = (1/n)(1 - 1/n)^{k-1}$, for $k = 1, 2, 3, \ldots$. Show that X_n/n converges to the exponential distribution with mean 1 in distribution.

10.10. (Slutsky Theorem) Suppose $X_n \xrightarrow{D} X$ and $Y_n \xrightarrow{D} c$ for some constant c. Assume that X_n and Y_n are defined on the same probability space for each n. Prove (a) $X_n + Y_n \xrightarrow{D} X + c$, and (b) $X_n Y_n \xrightarrow{D} cX$. (Hint: Use Exercise 10.3 and proceed as in the proof of Theorem 10.7.)

Laws of Large Numbers

11

The laws of large numbers give operational meaning to the expectation of a random variable. If we generate independent realizations of a random variable and compute their mean, the weak law of large numbers states that the sample mean converges to the expected value in probability as we increase the number of samples. In this chapter, we present two versions of the weak law of large numbers. The first one assumes that the random variables have finite variance, while the second one assumes that the mean is finite.

The weak law has numerous applications. We discuss two sample applications: Monte Carlo integration and data compression. Monte Carlo integration is a simple but effective method for approximating integral in high dimensions, which arises in financial computations. Data compression is a topic in information theory. The fundamental result by Shannon on data compression is based on the weak law of large numbers. We end this chapter with a version of the strong law of large number that assumes the fourth moments of the random variables are finite.

11.1 Some Useful Bounds and Inequalities

The first inequality is the complex version of Cauchy–Schwarz inequality. In the next theorem, X^* denote the complex conjugate of X.

Theorem 11.1 (Cauchy–Schwarz Inequality)
For complex-valued random variables X and Y in L^2, the expectation $E[X^*Y]$ is well-defined, and

$$|E[X^*Y]| \leq \sqrt{E[|X|^2]E[|Y|^2]} \qquad (11.1)$$

(continued)

© The Author(s), under exclusive license to Springer Nature Switzerland AG 2023
K. Shum, *Measure-Theoretic Probability*, Compact Textbooks in Mathematics,
https://doi.org/10.1007/978-3-031-49830-5_11

Theorem 11.1 (continued)
holds, with equality holding iff X and Y are scalar multiple of each other with probability 1. When X and Y are real, $E[X^*Y]$ is simplified to $E[XY]$.

Proof Let X and Y be complex-valued random variables such that both $E[|X|^2]$ and $E[|Y|^2]$ are finite. We first show that $|X^*Y|$ is integrable. Let A_1 be the subset of Ω in which $|X(\omega)| > |Y(\omega)|$, and A_2 be the subset in which $|X(\omega)| \leq |Y(\omega)|$. Then, we have

$$E[|X^*Y|] = \int_{A_1} |X^*Y| + \int_{A_2} |X^*Y| \leq \int_{A_1} |X|^2 + \int_{A_2} |Y|^2 \leq E[|X|^2] + E[|Y|^2].$$

This shows that $\int |X^*Y|$ is finite, and hence, X^*Y is integrable.

If $E[|X|^2] = 0$, then $X = 0$ almost surely, and hence, $X^*Y = 0$ almost surely. Both sides of (11.1) are equal to zero. Suppose $E[|X|^2] > 0$. Define a function $f(z) \triangleq E[|zX + Y|^2]$ with z being a complex variable. We can expand it as

$$f(z) = E[(zX + Y)^*(zX + Y)] = |z|^2 E[|X|^2] + z^* E[X^*Y] + z E[XY^*] + E[|Y|^2].$$

By completing square, we obtain

$$f(z) = E[|X|^2] \cdot \left| z + \frac{E[X^*Y]}{E[|X|^2]} \right|^2 - \frac{|E[X^*Y]|^2}{E[|X|^2]} + E[|Y|^2].$$

Since $f(z) \geq 0$ for all z, the constant term on the right must be nonnegative. This yields the Cauchy–Schwarz inequality.

When equality holds, then $f(z_0) = 0$ for a particular choice of z_0. This implies that $E[|z_0 X + Y|^2] = 0$, and hence, $Y = -z_0 X$ almost surely. ∎

Note that the proof of Theorem 11.1 does not rely on the assumption that the underlying measure space is a probability space. The Cauchy–Schwarz inequality can be derived for a general measure, as long as X and Y are square integrable.

Theorem 11.2 (Chebyshev Inequality)
If the variance of a real-valued random variable X is finite, then for any $\epsilon > 0$,

$$P(|X - E[X]| \geq \epsilon) \leq \frac{\text{Var}(X)}{\epsilon^2}.$$

Proof The indicator function $\mathbf{1}_{\{|X-E[X]|\geq\epsilon\}}(\omega)$ satisfies

$$\mathbf{1}_{\{|X-E[X]|\geq\epsilon\}}(\omega) \leq (X(\omega) - E[X])^2/\epsilon^2 \quad \text{for all } \omega \in \Omega.$$

Taking the expectation of both sides, we obtain

$$P(|X - E[X]| \geq \epsilon) = E[\mathbf{1}_{\{|X-E[X]|\geq\epsilon\}}] \leq \frac{E[(X - E[X])^2]}{\epsilon^2} = \frac{\text{Var}(X)}{\epsilon^2}.$$

∎

The next result is called *Jensen inequality*.

> **Theorem 11.3 (Jensen Inequality)**
> *Suppose ϕ is a convex function and X is a real-valued random variable with finite mean. Then $E[\phi(X)] \geq \phi(E[X])$.*

Proof By the convexity of ϕ, at each point x^* in the domain of ϕ, we can draw a straight line that touches the graph of ϕ at one point, i.e., it passes through the point $(x^*, \phi(x^*))$ and lies beneath the graph of $\phi(x)$.

We pick $x^* = E[X]$, which is finite by assumption. Let $g(x) = ax + b$ be a linear function such that $g(E[X]) = \phi(E[X])$ and $g(x) \leq \phi(x)$ for all x. If ϕ is not differentiable at $x^* = E[X]$, there could be more than one choice of coefficients a and b. In this case, we just pick one arbitrarily.

By applying the monotonic property of integral, we obtain

$$E[\phi(X)] \geq E[g(X)] = aE[X] + bE[1] = aE[X] + b = \phi(E[X]).$$

Note that we have used $E[1] = 1$ in the second last equality. ∎

11.2 Weak Law of Large Numbers

In the followings, we consider a sequence of independent random variables X_1, X_2, $X_3, \ldots$ defined on a common probability space $(\Omega, \mathcal{F}, P)$. For $n \geq 1$, denote the sum of the first n random variables by

$$S_n \triangleq X_1 + X_2 + \cdots + X_n.$$

We recall the following basic property of variance for pairwise uncorrelated random variables (see Definition 7.3).

Theorem 11.4
If $X_1, X_2, \ldots, X_n$ are pairwise uncorrelated, then

$$\mathrm{Var}(X_1 + X_2 + \cdots + X_n) = \sum_{i=1}^{n} \mathrm{Var}(X_i).$$

Proof Without loss of generality, suppose the random variables $X_1, \ldots, X_n$ have zero mean. (We can consider the centered random variable $Y_n = X_n - E[X_n]$ otherwise.)

Using the definition of variance and the linearity of expectation, we have

$$\mathrm{Var}(X_1 + X_2 + \cdots + X_n) = E\big[X_1^2 + X_2^2 + \cdots + X_n^2 + 2\sum_{i<j} X_i X_j\big] = \sum_{i=1}^{n} E[X_i^2].$$

The assumption of pairwise uncorrelatedness guarantees that the cross-terms $E[X_i X_j]$ are zero for $i \neq j$. Therefore, we have

$$\mathrm{Var}(X_1 + X_2 + \cdots + X_n) = \sum_{i=1}^{n} E[X_i^2] = \sum_{i=1}^{n} \mathrm{Var}(X_i).$$

∎

We will present two versions of weak law of large numbers. The first one assumes that the random variables are pairwise uncorrelated with the same mean and variance. Note that the random variables need not be identically distributed, and they need not be independent.

Theorem 11.5 (Weak Law of Large Numbers (L^2 Version))
Let $(X_i)_{i=1}^{\infty}$ be a sequence of pairwise uncorrelated random variables with common mean $E[X_i] = \mu$ and variance $\mathrm{Var}(X_i) = \sigma^2$ for all i. Then $S_n/n \xrightarrow{P} \mu$.

Proof The proof is an application of Chebyshev inequality (Theorem 11.2).

Fix any $\epsilon > 0$ and integer n. Using the fact that S_n/n is an average of pairwise uncorrelated random variables with common mean μ, we have

$$P\left(\left|\frac{S_n}{n} - \mu\right| > \epsilon\right) = P\left(\left|S_n - n\mu\right| > n\epsilon\right) \le \frac{\text{Var}(S_n)}{n^2\epsilon^2} = \frac{1}{n^2\epsilon^2}\sum_{i=1}^{n}\text{Var}(X_i).$$

The last equality follows from Theorem 11.4.

Using Theorem 11.4, we know that $\sum_{i=1}^{n}\text{Var}(X_i) = n\sigma^2$. Therefore, we have

$$P\left(\left|\frac{S_n}{n} - \mu\right| > \epsilon\right) \le \frac{n\sigma^2}{n^2\epsilon^2} = \frac{\sigma^2}{n\epsilon^2}.$$

As $n \to \infty$, the right-hand side approaches zero, which implies that S_n/n converges in probability to μ. This completes the proof. ∎

The assumption of finite second moment can be relaxed to finite mean, but at the cost of strengthening the uncorrelated assumption to independence.

Theorem 11.6 (Weak Law of Large Numbers (L^1 Version))
Suppose $(X_i)_{i=1}^{\infty}$ is a sequence of i.i.d. random variables with finite mean μ.
Then $S_n/n \xrightarrow{P} \mu$.

Theorem 11.5, also known as Khinchin's weak law of large numbers, has a detailed proof in [4, Theorem 2.2.14]. In the following two sections, we will discuss two applications of this theorem instead of going through the proof.

11.3 Application: Monte Carlo Integration

Suppose we have a multi-variable function $g(x_1, x_2, \ldots, x_d)$ with d variables, where each variable takes a value between 0 and 1. Our goal is to evaluate the integral

$$\int_0^1 \int_0^1 \cdots \int_0^1 g(x_1, x_2, \ldots, x_d)\, dx_1 dx_2 \cdots dx_d, \tag{11.2}$$

which is a d-fold integration over the unit hypercube. A direct method that approximates this integral by a Riemann sum involves selecting m points along each dimension and computing the sum

$$\frac{1}{m^d} \sum_{j_1=0}^{m-1} \sum_{j_2=0}^{m-1} \cdots \sum_{j_d=0}^{m-1} g\left(\frac{j_1}{m}, \frac{j_2}{m}, \ldots, \frac{j_d}{m}\right).$$

However, this method requires evaluating the function at m^d points, which can be computationally infeasible when d is large. This is known as the "curse of dimensionality", where even a small number of sample points per dimension can result in an exponential increase in the total number of function evaluations required.

To overcome this limitation, we can use Monte Carlo integration, which involves randomly sampling points from the domain of the hypercube $[0, 1]^d$, and taking the average of the function values at the random points. To estimate the integral using n sample points, we generate independent uniform random variables X_{i1}, $X_{i2}, \ldots, X_{id}$ between 0 and 1, for $i = 1, 2, \ldots, n$. We then compute the estimator

$$I_n \triangleq \frac{1}{n} \sum_{i=1}^{n} g(X_{i1}, X_{i2}, \ldots, X_{id}). \tag{11.3}$$

The n terms in this estimator are independent, and its expected value is given by

$$E[I_n] = \frac{1}{n} \sum_{i=1}^{n} E[g(X_{i1}, X_{i2}, \ldots, X_{id})] = E[g(X_{11}, X_{12}, \ldots, X_{1d})],$$

which is equal to the multi-variable integral in (11.2). If the integral in (11.2) is finite, then by Khinchin's weak law of large numbers, the estimator I_n converges in probability to its expected value $E[I_n]$ as $n \to \infty$.

The convergence rate of the Monte Carlo estimator is independent of the dimensionality of the problem. To see why, suppose that the function g to be integrated is bounded by a constant C in absolute value, and suppose the error tolerance is ϵ. Let the true value of the integral be I^*. Because the variance is finite, by the L^2 version of the weak law of large numbers, we have

$$P\left(\left|I_n - I^*\right| > \epsilon\right) \le \frac{\mathrm{Var}(g(X_{11}, \ldots, X_{1d}))}{n\epsilon^2} \le \frac{C^2}{n\epsilon^2}.$$

If we want the probability $P(|I_n - I^*| > \epsilon)$ to be less than, say, 0.01, we need to take at least $C^2/(0.01\epsilon^2)$ samples. The bound increases inversely proportional to ϵ^2. This error estimate does not depend on the number of variables and is valid in any dimension. This means that the convergence rate of the Monte Carlo estimator is the same for one-dimensional problems as it is for high-dimensional problems. However, the variance of the estimator can still be high for functions with large fluctuations, requiring the evaluation of the function on many sample points to achieve a certain level of accuracy.

We illustrate the convergence speed of Monte Carlo integration with the following example.

▶ **Python Program for Monte Carlo Integration** We demonstrate the Monte Carlo method by evaluating the integral

$$\int_{-M}^{M} \int_{-M}^{M} \int_{-M}^{M} \int_{-M}^{M} x^2 y^2 z^2 w^2 \, dx dy dz dw$$

with four variables. The integral is separable, with exact value $(2M^3/3)^4$. We can implement the Monte Carlo integration method using the following Python program:

```
from random import uniform
n = 100000                                   # number of sample points
f = lambda u:  (u[0]*u[1]*u[2]*u[3])**2    # function to be integrated
M = 2                                        # range of the integration

s = 0
for i in range(n):
    random_point = [uniform(-M,M) for j in range(4)]
    s += f(random_point)                             # evaluate function f

print("Integral = ", (2*M)**4*s/n)
```

We set $M = 2$. The exact value of the integral is 809.086. The Python function uniform($-M,M$) generates a float-type number uniformly between $-M$ and M. We run the program with $n = 100, 10^3, 10^4, 10^5$, and 10^6. The approximate values produced by the program are shown in the table below.

n	100	1000	10000	100000	1000000
I_n	724.062	822.986	796.259	812.225	807.883

When n is large, it is close to the exact value 809.086.

Quasi-Random Monte Carlo

Another method for numerical evaluation of multi-variable is to use *quasi-random point set* instead of pseudo-random numbers. Quasi-random point sets are sequence of points that are evenly spread in the hypercube $[0, 1]^d$ but are not statistically the same as uniformly random points. These sequences are also called "low-discrepancy sequences". If we use quasi-random point sets in Monte Carlo integration, the estimate can converge to the true value of the integral faster than random points, especially in high-dimensional problems.

11.4   Application: Data Compression

Let $\mathcal{A} = \{a_1, a_2, \ldots, a_m\}$ denote an alphabet of size m. Suppose we generate a string of length n, with each character drawn independently according to a probability distribution $P(a_i) = p_i$, where p_i's are nonnegative real numbers summing to 1. There are totally m^n combinations. We may associate each string with an integer between 1 and m^n. Using $\log_2(m^n)$ bits, we can store and recover the random string with no decoding error.

However, if we allow an arbitrarily small error in decoding, we can significantly reduce the number of bits. It turns out there exists a set $S \subset \mathcal{A}^n$ with probability $\Pr(S) \geq 1 - \epsilon$ and $|S| \ll m^n$. We only encode the strings in the set S. The strings not in S are not encoded, or encoded arbitrarily. The decoding error is thus less than or equal to ϵ. We will show that ϵ can be made arbitrarily small as the block length n increases, with the cardinality of the set S roughly equal to $2^{nH(P)}$, where $H(P)$ denotes the entropy of the distribution P,

$$H(P) \triangleq - \sum_{i=1}^{m} p_i \log_2 p_i.$$

To realize this idea, we fix a parameter $\delta > 0$ and define a set $S_{\delta,n}$ of strings by

$$S_{\delta,n} \triangleq \{(x_1, x_2, \ldots, x_n) \in \mathcal{A}^n : 2^{-n(H(P)+\delta)} \leq \prod_{k=1}^{n} p(x_k) \leq 2^{-n(H(P)-\delta)}\}.$$

We claim that for any fixed $\delta > 0$, the set $S_{\delta,n}$ has probability arbitrarily close to 1 as $n \to \infty$. We denote the random string of length n by $\mathbf{X}$, and for $k = 1, 2, \ldots, n$, let X_k be the k-th letter in $\mathbf{X}$. Define a new random variable $U_k = -\log_2 P(X_k)$, where $P(X_k)$ is equal to p_i if $X_k = i$. For simplicity, we may assume $p_i > 0$ for all i, so that $\log_2 p_i$ is well-defined. The expectation of U_k is precisely equal to the entropy $H(P)$ of distribution P. By the weak law of large numbers, for any fixed $\delta > 0$, we have

$$\lim_{n \to \infty} \Pr\left(\left|\frac{1}{n}\sum_{k=1}^{n} U_k - H(P)\right| \leq \delta\right) = 1.$$

But

$$\left|\frac{1}{n}\sum_{k=1}^{n} U_k - H(P)\right| \leq \delta \Leftrightarrow 2^{n(H(P)-\delta)} \leq 2^{U_1+U_2+\cdots+U_n} \leq 2^{n(H(P)+\delta)}$$

$$\Leftrightarrow 2^{-n(H(P)+\delta)} \leq P(X_1)P(X_2)\cdots P(X_n) \leq 2^{-n(H(P)-\delta)}.$$

The random string $\mathbf{X}$ belongs to $S_{\delta,n}$ if and only if $\left|\frac{1}{n}\sum_{k=1}^{n} U_k - H(P)\right| \leq \delta$. This shows that the probability of the set of strings $S_{\delta,n}$ approaches 1 as $n \to \infty$, for any fixed $\delta > 0$. To estimate the cardinality of $S_{\delta,n}$, we note that all random strings in $S_{\delta,n}$ have probability at least $2^{-n(H(P)+\delta)}$. There are at most $2^{n(H(P)+\delta)}$ strings in $S_{\delta,n}$. We can represent them using at most $n(H(P) + \delta)$ bits.

To design the data compression scheme with decoded error no more than ϵ, we first fix a small $\delta > 0$ and find a sufficiently large block length n such that $\Pr(\mathbf{X} \in$

$S_{\delta,n}) > 1 - \epsilon$. Because we only encode the strings in $S_{\delta,n}$, it takes $H(P) + \delta$ bits per symbol. The saving is particularly significant if $H(P) \ll \log_2(m)$.

The result demonstrated in this example is the forward part of Shannon's theorem on data compression. This is one of the first major results in information theory [3].

We demonstrate the idea of this data compression with finite block length n. Suppose that alphabet is $\mathcal{A} = \{a, b, c\}$. The data source generates symbols a, b, and c independently, according to the distribution $P(a) = 2/3$, $P(b) = 2/9$ and $P(c) = 1/9$. We group the random symbols into groups of size $n = 11$ and use B bits to encode each group. Only the $2^B - 1$ strings of length n with the largest probabilities are encoded. If the random string is not among these $2^B - 1$ selected ones, we use a special number, such as 2^B, to signify that a decoding error has occurred.

The probability of error is calculated by the following Python program.

▶ **Python Program for Demonstrating Data Compression**

```python
# Compute the success decoding probability of
# data compression with fixed block length
from numpy import prod
from itertools import product

n = 11  # block length
prob_dist= {'a':2/3, 'b':2/9, 'c':1/9}
B = 16

A = [x for x in prob_dist]  # list of letters
D={}
for t in product(A, repeat = n):
  s = ''.join(t)
  D[s] = prod([prob_dist[x] for x in t])

# sum the largest 2**B-1 probabilities
list_of_prob = list(D.values())
list_of_prob.sort(reverse = True)
prob_success = sum(list_of_prob[0:(2**B-1)])
print(f"{B} bits per block")
print(f"P(successful decoding) = {prob_success}")
```

The table below shows the probabilities of successful decoding for a different number of bits per block. The probabilities are obtained by running the Python program for $B = 8, 9, \ldots, 16$.

No. of bits per block	8	9	10	11	12	13	14	15	16
P (successful decoding)	0.245	0.327	0.428	0.533	0.650	0.765	0.864	0.939	0.983

We observe that the value of $11 \cdot H(P) = 11(1.22) \approx 13.475$, where $H(P) \approx 1.22$ is entropy of the probability distribution P. The table shows that there is a significant increase in the probability of successful decoding when the number of bits per block is slightly above the threshold value, i.e., for $B = 14$ or higher.

In general, as the block length n approaches infinity, there is an abrupt increase in success probability when the number of bits per block is $1.22n$.

11.5 Strong Law of Large Numbers

If we assume that the fourth moments of the random variables are bounded, we have an easy proof of the strong law of large number.

> **Theorem 11.7 (4th-moment Strong Law of Large Numbers)**
> *Suppose X_i, for $i \geq 1$, are independent random variables with mean μ and $E[X_i^4] \leq c < \infty$ for all i. Then $\frac{S_n}{n} \to \mu$ almost surely.*

Proof Without loss of generality, we assume that $\mu = 0$. (We can consider $Y_i = X_i - \mu$ if the mean of X_i is not zero.)

The expectation of S_n^4 can be expanded as

$$E[S_n^4] = E[(X_1 + X_2 + \cdots + X_n)^4]$$

$$= E\Big[\sum_{i=1}^{n} X_i^4 + 3 \sum_{i \neq j} X_i^2 X_j^2 + 4 \sum_{i \neq j} X_i X_j^3$$

$$+ 6 \sum_{\substack{i,j,k \\ i,j,k \text{ distinct}}} X_i X_j X_k^2 + \sum_{\substack{i,j,k,\ell \\ i,j,k,\ell \text{ distinct}}} X_i X_j X_k X_\ell \Big].$$

We analyze the terms one by one. For distinct indices i, j, k, and ℓ, we have

$$E[X_i X_j X_k X_\ell] = E[X_i X_j X_k^2] = E[X_i X_j^3] = 0,$$

by the assumption that $X_1, \ldots, X_n$ are independent. The fourth-power terms are bounded by $E[\sum_{i=1}^{n} X_i^4] \leq nc$ because $E[X_i^4] \leq c$.

For $i \neq j$, by Cauchy–Schwarz inequality,

$$E[X_i^2 X_j^2] \leq \sqrt{E[X_i^4] E[X_j^4]} \leq \sqrt{c \cdot c} = c.$$

Hence,

$$3E\Big[\sum_{i \neq j} X_i^2 X_j^2 \Big] \leq 3n(n-1)c$$

This gives $E[S_n^4] \le nc + 3n(n-1)c$. We now fix any $\epsilon > 0$,

$$P\left(\frac{|S_n|}{n} > \epsilon\right) = P(S_n^4 > n^4\epsilon^4) \le \frac{nc + 3n(n-1)c}{n^4\epsilon^4} \le \frac{3n^2c + nc - 3nc}{n^4\epsilon^4} \le \frac{3c}{n^2\epsilon^4}.$$

Because

$$\sum_{n=1}^{\infty} P\left(\frac{|S_n|}{n} > \epsilon\right) \le \sum_{n=1}^{\infty} \frac{3c}{n^2\epsilon^4} < \infty,$$

we can apply the first Borel–Cantelli lemma (Theorem 5.8) to conclude that the event $\{|S_n|/n > \epsilon \text{ i.o.}\}$ has probability 0. Therefore, by Theorem 10.1, S_n/n converges to 0 almost surely.   ∎

Note that Theorem 11.7 does not assume that the random variables X_i are identically distributed, but it requires that the 4th moments are uniformly bounded. We state below a stronger version the strong law that assumes finite mean and pairwise independence.

> **Theorem 11.8**
> *Suppose X_i, for $i \ge 1$, are pairwise independent and identically distributed random variables with finite mean μ. Then $S_n/n \xrightarrow{a.s.} \mu$ as $n \to \infty$.*

A proof of this theorem by Etemadi can be found in [4, Thm 2.4.1].

Example 11.5.1 (Estimation of Cumulative Distribution Function)
Consider the problem of estimating the cdf $F(x) = \Pr(X \le x)$ of a random variable X by generating i.i.d. samples X_n from the cdf F for $n \ge 1$. We can estimate $F(x)$ by the empirical distribution

$$\hat{F}_n(x) \triangleq \frac{1}{n} \sum_{k=1}^{n} \mathbf{1}_{[X_k,\infty)}(x).$$

The indicator function $\mathbf{1}_{[X_k,\infty)}(x)$ is equal to 1 if $x \ge X_k$ and is 0 otherwise. The function $\hat{F}_n$ first counts the number of samples that is less than or equal to x, and then divides it by n. Because it depends on the random samples, the estimate $\hat{F}_n(x)$ is a random variable. By the strong law of large numbers, $\hat{F}_n(x)$ converges to the correct value $F(x)$ with probability 1.
This result is strengthened in the Glivenko–Cantelli theorem, which states that the convergence is uniform in x,

$$\sup_{x \in \mathbb{R}} |\hat{F}_n(x) - F(x)| \xrightarrow{a.s.} 0.$$

The Glivenko–Cantelli theorem is informally known as "the fundamental theorem of statistics".

Example 11.5.2 (Normal Numbers)
A real number is said to be *normal* in base 10 if the frequency of each digit 0 through 9 in its decimal expansion is asymptotically equal. If a number has multiple decimal expansions, such as $0.1 = 0.09999\ldots$, we choose the one that does not end in infinitely many trailing 0's. The strong law of large numbers implies that, with probability 1, a randomly chosen number in the interval $[0, 1]$ is normal in base 10.

We can generalize this notion of normality to other bases. A real number is said to be *normal* if it is normal in base b all for all integer $b \geq 2$ simultaneously. By the strong law of large numbers, for any integer $b \geq 2$, there exists a set A_b in $[0, 1]$ with $P(A_b) = 1$ such that all numbers in A_b are normal in base b. Taking the intersection of countably events with probability 1, we deduce that a number in $[0, 1]$ is normal almost surely. This proves the existence of normal numbers.

However, the argument in the previous paragraph is non-constructive. It is a non-trivial task to explicitly construct a normal number or prove that a given number is normal.

Problems

11.1. Suppose $r < s$. Show that if $E[|X|^s]$ is finite, then $E[|X|^r]$ is also finite.

11.2. (Cantelli inequality) Let X be a random variable with finite variance, and let $a \geq 0$. Show that

$$P(X - E[X] \geq a) \leq \frac{\text{Var}(X)}{\text{Var}(X) + a^2}.$$

11.3. Let $X_1, X_2, \ldots, X_n$ be i.i.d. random variables uniformly distributed between 0 and 1. Show that the sequence of random variables

$$Y_n \triangleq \frac{X_1 + X_2 + \cdots + X_n}{X_1^2 + X_2^2 + \cdots + X_n^2}$$

converges to a constant in probability as $n \to \infty$.

11.4. Let $(X_n)_{n \geq 1}$ be i.i.d. random variables with pdf $f(x) = 2x^{-3}\mathbf{1}_{[1,\infty)}(x)$. Apply Khinchin's weak law to show that $(X_1 + X_2 + \cdots + X_n)/n \xrightarrow{P} 2$.

11.5. Suppose that X_k, for $k = 1, 2, 3, \ldots$, are independent random variables with zero mean and variance $\text{Var}(X_k) \leq c\sqrt{k}$ for some positive constant c. Prove that X_n/n converges almost surely to zero as $n \to \infty$.

11.6. Suppose $(X_k)_{k=1}^{\infty}$ is a sequence of i.i.d. random variables with zero mean, and suppose the random variables are uniformly bounded, i.e., there is a constant c such that $P(|X_k| < c) = 1$ for all k. Let $S_n = X_1 + \cdots + X_n$. Prove that $S_n/n^{3/4} \xrightarrow{a.s.} 0$ as $n \to \infty$.
(Hint: Show that $E[S_n^6]$ is upper bounded by cn^3 for some constant c.)

11.7. Let $(X_i)_{i=1}^\infty$ be a sequence of random variables that may be correlated. Suppose $E[X_i] = \mu$ and $\mathrm{Var}(X_i) \le K$ for all i. Show that if there exists a sequence of real numbers $(a_\tau)_{\tau \ge 1}$ such that (i) $a_\tau \in [0, 1]$ for all τ, (ii) a_τ decreases to 0 as τ increases, and (iii) $\mathrm{Cov}(X_i, X_{i+\tau}) \le a_\tau \sqrt{\mathrm{Var}(X_i)\,\mathrm{Var}(X_{i+\tau})}$ for all i and $\tau > 0$, then $(X_i)_{i=1}^\infty$ converges in probability to μ.

11.8. Define a first-order autocorrelation process by $X_0 = 0$ and $X_i = \rho X_{i-1} + N_i$ for $i \ge 1$, where ρ is a constant with $|\rho| < 1$ and N_i is a Gaussian random variable $\sim N(0, \sigma^2)$. Assume that the random variables N_i's are independent. Show that $(X_1 + X_2 + \cdots + X_i)/i$ converges to 0 in probability as $i \to \infty$.

11.9. Consider the experiment of throwing k distinct balls into n distinct boxes. Assume the locations of the balls are independent and uniformly chosen among $\{1, 2, \ldots, n\}$. We consider an asymptotic scenario in which k and n increase simultaneously with $k/n \to a$, for some positive constant a. Let X_n denote the number of empty boxes when the number of boxes is n. Prove that $X_n/n \to e^{-a}$ in quadratic mean, and hence in probability as $n \to \infty$.

11.10. Prove Glivenko–Cantelli theorem under a simplifying assumption that the limit cdf $F(x)$ is continuous.

Techniques from Hilbert Space Theory

12

This chapter presents a geometric perspective on random variables by viewing them as vectors in a vector space. Although this vector space has infinite dimension in general, this viewpoint offers insights into the behavior of random variables. Assuming that all random variables in the vector space have finite second moments, we can define an inner product that resembles the dot product in a finite-dimensional Euclidean space. This inner product allows us to define notions such as orthogonality and the analog of the triangle inequality in this infinite-dimensional vector space.

To study random variables form this geometric perspective, we borrow techniques from Hilbert space theory. We can define the projection operator, which maps a random variable to a subspace, and prove the Orthogonality Principle in the context of probability theory. These tools are then applied to the derivation of statistical estimators that minimize mean-squared error.

12.1 L^2-Norm and Inner Product Space

Definition 12.1

Given a measure space $(\Omega, \mathscr{F}, \mu)$, we denote the set of all $\mathscr{F}$-measurable functions X with finite second moment by $L^2(\mu)$. More generally, for any positive integer p, we define $L^p(\mu)$ as the set of all $\mathscr{F}$-measurable functions with finite $\int |X|^p \, d\mu$.

In this chapter, we will focus on the study of $L^2(P)$ for a probability measure P. We note that any square-integrable random variable necessarily has a finite mean, implying that $L^2(P)$ is a subset of $L^1(P)$. Using Cauchy–Schwarz inequality (Theorem 11.1), we can define an inner product for two square-integrable random variables.

© The Author(s), under exclusive license to Springer Nature Switzerland AG 2023
K. Shum, *Measure-Theoretic Probability*, Compact Textbooks in Mathematics,
https://doi.org/10.1007/978-3-031-49830-5_12

Definition 12.2

Let $(\Omega, \mathscr{F}, P)$ be a probability space. We define the *inner product* of two real-valued random variables in $L^2(P)$ by

$$\langle X, Y \rangle \triangleq E[XY].$$

We define the L^2-*norm* of X by

$$\|X\|_2 \triangleq \sqrt{\langle X, X \rangle} = \sqrt{E[X^2]}.$$

Two random variables X and Y are said to be *orthogonal* if $\langle X, Y \rangle = 0$. We will sometime write $\|X\|$ without the subscript $_2$ to simplify notation.

This notion of orthogonality is analogous to the concept of perpendicularity in Euclidean geometry. When random variables X and Y have zero mean, then they are orthogonal iff they are uncorrelated.

▶ **Complex Inner Product** For complex random variables, the inner product is defined as $E[X^*Y]$, where X^* denotes the complex conjugate of X.

Example 12.1.1 (A Finite Inner Product Space)
Consider a finite probability space with sample space $\Omega = \{a, b, c\}$. The σ-algebra is the discrete σ-algebra $\mathscr{P}(\Omega)$, and the probability measure is given by $P(\{a\}) = p_a$, $P(\{b\}) = p_b$, $P(\{c\}) = p_c$, where p_a, p_b, and p_c are nonnegative real numbers that sum to 1.

We can specify a random variable X by giving its value at the three points in Ω, which can be represented as a vector $(X(a), X(b), X(c))$. Conversely, any three-dimensional vector (x_1, x_2, x_3) is associated to a random variable X by defining $X(a) \triangleq x_1$, $X(b) \triangleq x_2$, and $X(c) \triangleq x_3$. Thus, the random variables in this probability space have a one-to-one correspondence with the vectors in a three-dimensional vector space over $\mathbb{R}$.

The inner product between two random variables X and Y, which are represented by vectors (x_1, x_2, x_3) and (y_1, y_2, y_3), respectively, is computed as

$$\langle X, Y \rangle = p_a x_1 y_1 + p_b x_2 y_2 + p_c x_3 y_3.$$

The second moment of a random variable X is thus equal to $p_a x_1^2 + p_b x_2^2 + p_b x_3^2$.

Example 12.1.2 (A Complex Hilbert Space of Infinite Dimension)
Let $\Omega = \{1, 2, 3, \ldots\}$ be the sample space, and the power set $\mathscr{P}(\Omega)$ be the σ-algebra. Consider a probability measure P defined on Ω by

$$P(\{k\}) \triangleq \frac{1}{2^k}$$

for $k \geq 1$. A random variable X can be identified with an infinite complex sequence

$$X \leftrightarrow (x_1, x_2, x_3, \ldots),$$

where x_k are complex numbers for $k = 1, 2, 3, \ldots$.

The random variable X belongs to $L^2(P)$ if and only if the series $\sum_{k=1}^{\infty} |x_k|^2 / 2^k$ is convergent. If we are given another random variable Y, whose values at $1, 2, 3, 4, \ldots$ are $y_1, y_2, y_3, y_4, \ldots$, the inner product between X and Y is given by

$$\langle X, Y \rangle = \sum_{k=1}^{\infty} x_k^* y_k / 2^k.$$

Here, x_k^* denotes the complex conjugate of x_k.

The followings are other basic properties of inner product.

Theorem 12.1
Let $\langle \cdot, \cdot \rangle$ be an inner product over a field K, where K is either $\mathbb{R}$ or $\mathbb{C}$:

- *For $X, Y_1, Y_2 \in L^2(P)$ and $\alpha, \beta \in K$,*

$$\langle X, \alpha Y_1 + \beta Y_2 \rangle = \alpha \langle X, Y_1 \rangle + \beta \langle X, Y_2 \rangle.$$

- *For real inner product, we have $\langle X, Y \rangle = \langle Y, X \rangle$ for $X, Y \in L^2(P)$.*
- *For complex inner product, we have $\langle X, Y \rangle = \langle Y, X \rangle^*$ for $X, Y \in L^2(P)$.*
- *For $X \in L^2(P)$ and $\alpha \in K$, $\|\alpha X\|_2 = |\alpha| \cdot \|X\|_2$.*

The triangle inequality for L^2 norm is called the Minkowski inequality.

Theorem 12.2 (Minkowski Inequality)
For any two random variables X and Y in $L^2(P)$,

$$\|X + Y\|_2 \leq \|X\|_2 + \|Y\|_2.$$

Proof We give a proof for complex-valued random variables. Using the Cauchy–Schwarz inequality (Theorem 11.1),

$$\|X + Y\|_2^2 \leq E[|X|^2] + 2\operatorname{Re}(E[X^* Y]) + E[|Y|^2]$$

$$\leq E[|X|^2] + 2\sqrt{E[|X|^2] E[|Y|^2]} + E[|Y|^2]$$

$$= (\|X\|_2 + \|Y\|_2)^2.$$

Taking the square roots of both sides yields the Minkowski inequality. ∎

Similar to the case of L^1 norm, a random vector with zero L^2 norm is equivalent to the zero function but need not be equal to the zero function. We continue with the convention that two random variables that are equal almost everywhere are regarded as the same random variable. With this caveat, we have shown that $L^2(P)$ is a normed vector space.

Using the Minkowski inequality, we can see that the L^2 norm defines a metric on the space $L^2(P)$, by regarding $\|X - Y\|_2$ as the distance between two random variables. Moreover, we can prove that the space of random variables in $L^2(P)$ is complete with respect to the norm induced by inner product. It is a consequence of the Riesz–Fischer theorem.

Definition 12.3

A sequence of vectors $(\mathbf{u}_n)_{n=1}^{\infty}$ in a vector space V with a norm function $\| \cdot \|$ is called a *Cauchy sequence* if given any $\epsilon > 0$, there exists an integer N such that $\|\mathbf{u}_m - \mathbf{u}_n\| \leq \epsilon$ for all $m, n \geq N$. We say that a sequence of vectors $(\mathbf{u}_n)_{n=1}^{\infty}$ is *convergent* if there exists a vector $\mathbf{u}_0 \in V$ such that for any given $\epsilon > 0$, there is an integer N such that $\|\mathbf{u}_n - \mathbf{u}_0\| \leq \epsilon$ for all $n \geq N$.

A normed vector space V is said to be *complete* if every Cauchy sequence in V is convergent. A *Hilbert space* is a (real or complex) vector space that is complete with respect to the norm induced by the inner product.

Theorem 12.3 (Riesz–Fischer)
Let $(\Omega, \mathscr{F}, P)$ be a probability space. The space $L^2(P)$ is a complete vector space with respect to the L^2 norm.

Proof Suppose we have a Cauchy sequence $f_1, f_2, f_3, \ldots$ in $L^2(P)$. To prove that the sequence converges in mean square, we need to find a candidate function and show that it is indeed the limit. Our first step is to show that there exists a subsequence that converges almost surely.

By the definition of Cauchy sequence, we can find a sequence of integers $n_1 < n_2 < n_3 < \cdots$ such that $\|f_{n_{k+1}} - f_{n_k}\|_2 \leq 1/2^k$ for $k = 1, 2, 3, \ldots$. We claim that the limit $\lim_{k \to \infty} f_{n_k}(\omega)$ exists almost everywhere.

We write $f_{n_k}(\omega)$ as

$$f_{n_k}(\omega) = f_{n_1}(\omega) + \sum_{i=1}^{k-1}(f_{n_{i+1}}(\omega) - f_{n_i}(\omega))$$

and apply the result established in Problem 7.7, which implies that the limit of $f_{n_k}(\omega)$ exists almost everywhere if

$$\int |f_{n_1}|\, dP + \sum_{i=1}^{\infty} \int |f_{n_{i+1}} - f_{n_i}|\, dP \tag{12.1}$$

is finite.

To show that it is finite, we use Jensen inequality to obtain

$$\int |f_{n_{i+1}} - f_{n_i}|\, dP \le \left(\int |f_{n_{i+1}} - f_{n_i}|^2 \, dP \right)^{1/2} = \| f_{n_{i+1}} - f_{n_i} \|_2$$

for all i. Our choice of the indices n_i's ensures that $\| f_{n_{i+1}} - f_{n_i} \|_2 \le 1/2^i$. Therefore, the sum in (12.1) is finite. By applying the result from Problem 7.7, there exists an event E with probability 1 such that $f_{n_k}(\omega)$ converges for all $\omega \in E$. We can define a candidate function as follows:

$$f(\omega) \triangleq \begin{cases} \lim_{k \to \infty} f_{n_k}(\omega) & \text{for } \omega \in E, \\ 0 & \text{otherwise.} \end{cases}$$

If ω is not in E, the value of $f(\omega)$ is arbitrary. We set the value of $f(\omega)$ to be 0 for $\omega \in E^c$. This convention ensures that $f(\omega)$ is measurable.

Let $\epsilon > 0$ be an arbitrarily small real number. By the choice of the sequence n_j, there is a sufficiently large integer m such that $\| f_{n_j} - f_{n_k} \|_2^2 \le \epsilon$ for all $j, k \ge m$. We fix $k \ge m$ and let $j \to \infty$. Then, by applying Fatou's lemma (Theorem 7.3), we have

$$\int_E \lim_{j \to \infty} |f_{n_j} - f_{n_k}|^2 \, dP \le \lim_{j \to \infty} \int_E |f_{n_j} - f_{n_k}|^2 \, dP \le \epsilon.$$

Since $f_{n_j}(\omega) \to f(\omega)$ for $\omega \in E$, we get $\| f - f_{n_k} \|_2^2 = \int |f - f_{n_k}|^2 \, dP \le \epsilon$ for all $k \ge m$. Therefore, the subsequence $(f_{n_k})_{k \ge 1}$ converges to f in mean square.

To show that the original sequence $(f_n)_{n=1}^{\infty}$ also converges to f in mean square, we write

$$\| f_n - f \|_2 < \| f_n - f_{n_\ell} \|_2 + \| f_{n_\ell} - f \|_2,$$

and take limit as n and ℓ approach infinity. Since $(f_n)_{n \ge 1}$ is a Cauchy sequence, the term $\| f_n - f_{n_\ell} \|_2$ tends to zero. Furthermore, from the previous paragraph, we know that $\| f_{n_\ell} - f \|_2$ tends to zero as well. This proves that $\| f_n - f \|_2$ can be made arbitrarily small for all sufficiently large integer n, and thus $(f_n)_{n \ge 1}$ converges to f in mean square. Because the L^2 norm is a continuous function (Exercise 12.2), the L^2 norm of f is finite. This proves that $f \in L^2(P)$. ∎

12.2   Closed Subspace and Projection

We define the notion of closed subspace below.

Definition 12.4

A subset W of $L^2(P)$ is said to be:

- A *subspace* if $X, Y \in W$ implies $\alpha X + \beta Y \in W$ for all scalars α and β
- *Closed* if $X_n \in W$, for $n = 1, 2, 3, \ldots$, and $X_n \xrightarrow{L^2} X$, then $X \in W$

In any finite-dimensional vector space, such as $\mathbb{R}^n$, any subspace is closed. The assumption of closedness is important when we are dealing with vector space with infinite dimension.

Example 12.2.1 (An Example of Subspace that Is Not Closed)

In Example 12.1.2, we let W denote the subset of infinite sequences $(x_1, x_2, x_3, \ldots)$ in which only finitely many components are nonzero. This is a vector space because it is closed under addition and scalar multiplication.

However, this subspace is not closed because

$$X_k \triangleq \underbrace{(1, 1, 1, \ldots, 1}_{k \text{ ones}}, 0, 0, 0, \ldots,)$$

converges to an infinite sequence that contains infinitely many 1's, as $k \to \infty$, which is not in the subspace.

Example 12.2.2 (Simple Random Variables Associated with a Fixed Partition of the Sample Space)

Let $(\Omega, \mathscr{F}, P)$ be a probability space and let $A_1, A_2, \ldots, A_d$ be a partition of the sample space Ω. Let W denote the set consisting of random variables that are constant on each A_k, for $k = 1, 2, \ldots, d$. It is clear that W is a linear subspace of $L^2(P)$.

To show that W is closed, consider a Cauchy sequence $(X_n)_{n=1}^{\infty}$ in W. For each index $k = 1, 2, \ldots, d$, we can write $X_n(\omega)\mathbf{1}_{A_k}(\omega) = c_{nk}\mathbf{1}_{A_k}(\omega)$ for some constant c_{nk}. The sequence $(c_{nk})_{n=1}^{\infty}$ is a Cauchy sequence of real numbers and thus is converging to some constant, say γ_k, as $n \to \infty$.

Define a random variable X in W by $\sum_{k=1}^{d} \gamma_k \mathbf{1}_{A_k}$. Since $X_k \to X$ in $L^2(P)$, we see that W is a closed subspace.

Theorem 12.4   (Projection Theorem)

Given a closed subspace $W \subseteq L^2(P)$ and a random variable $Y \in L^2(P)$, there exists a unique random variable $X^* \in W$ such that

$$\|Y - X^*\|_2 = \inf_{X \in W} \|Y - X\|_2.$$

This theorem says that we can always find a point in W whose distance to Y is the smallest among all other points in W.

Proof (Existence) Suppose $\inf_{X \in W} \|Y - X\| = \alpha$. Since α is the largest lower bound of $\{\|Y - X\| : X \in W\}$, we can find a sequence $(\alpha_n)_{n=1}^{\infty}$ such that $\alpha_n \searrow \alpha$, such that for each n, α_n is the L^2 distance between Y and X_n for some random variable X_n in the subspace W. We claim that the sequence of random variables $(X_n)_{n=1}^{\infty}$ is a Cauchy sequence.

To prove this claim, we make use of the parallelogram law (Exercise 12.1), which says that for any two random variables U and V in $L^2(P)$,

$$\|U + V\|^2 + \|U - V\|^2 = 2\|U\|^2 + 2\|V\|^2.$$

Suppose m and n are two positive integers. In the parallelogram law, substitute U by $Y - X_m$ and V by $Y - X_n$. We have $U + V = 2Y - X_m - X_n$ and $U - V = X_n - X_m$. The parallelogram law gives

$$\|2Y - X_m - X_n\|^2 + \|X_n - X_m\|^2 = 2\|Y - X_m\|^2 + 2\|Y - X_n\|^2.$$

Arrange the terms to get

$$\|X_n - X_m\|^2 = 2\|Y - X_m\|^2 + 2\|Y - X_n\|^2 - 4\|Y - (X_n + X_m)/2\|^2.$$

In the last term, $(X_n + X_m)/2$ is the "mid-point" between X_n and X_m and is thus a random variable in W (because W is a subspace). The distance $\|Y - (X_n + X_m)/2\|$ cannot be less than α by the definition of infimum. Therefore,

$$\|X_n - X_m\|^2 \le 2\alpha_m^2 + 2\alpha_n^2 - 4\alpha^2.$$

Given any $\epsilon > 0$, we choose a positive integer N such that the right-hand side of the above inequality is less than ϵ for all $m, n \ge N$. This is possible because the sequence $(\alpha_k)_{k=1}^{\infty}$ is converging to α. This completes the proof of the claim that $(X_n)_{n=1}^{\infty}$ is a Cauchy sequence (with respect to the L^2 norm).

By the completeness of L^2 norm (Theorem 12.3), the sequence $(X_n)_{n=1}^{\infty}$ converges to a random variable X^* in $L^2(P)$. That is, there exists a random variable $X^* \in L^2(P)$ such that X_n converges to X^* in mean square. Since W is closed, this random variable X^* must be in W.

Finally, we can conclude that

$$\alpha = \inf_{X \in W} \|Y - X\| = \lim_{n \to \infty} \|Y - X_n\| = \|Y - X^*\|.$$

The last step is due to the continuity of the L^2 norm. This proves the existence part.

(Uniqueness) Suppose there are two random variables X_1^* and X_2^* in W such that

$$\|Y - X_1^*\| = \|Y - X_2^*\| = \inf_{X \in W} \|Y - X\| \triangleq \alpha.$$

We can apply the parallelogram law with $U = Y - X_1^*$ and $V = Y - X_2^*$ to obtain

$$\|X_1^* - X_2^*\|^2 \leq 2\|Y - X_1^*\|^2 + 2\|Y - X_2^*\|^2 - 4\alpha^2 = 2\alpha^2 + 2\alpha^2 - 4\alpha^2 = 0.$$

Therefore, $\|X_1^* - X_2^*\| = 0$. It can be true only when X_1^* and X_2^* are equal a.s. ∎

The infimum in Theorem 12.4 can indeed be achieved. The random variable $X^* \in W$ satisfies $\|Y - X^*\|_2 = \min_{X \in W} \|Y - X\|_2$. In view of the previous theorem, we make the following definition.

Definition 12.5

(Projection function) Given a closed subspace W in $L^2(P)$ and a random variable $Y \in L^2(P)$, the unique random variable X that minimizes $\|Y - X\|_2$ is called the *projection of Y onto W*. We denote this minimizer as

$$\text{Proj}_W(Y) \triangleq \arg\min_{X \in W} \|Y - X\|_2.$$

Many estimation problems can be formulated as projection onto a closed subspace. The basic linear regression is an example.

Example 12.2.3 (Linear Regression as Projection)

Given n data points (x_i, y_i), for $i = 1, 2, \ldots, n$, the problem of linear regression is to find coefficients a and b that minimize the mean-squared error $\sum_{i=1}^{n}(y_i - ax_i - b)^2$. To formulate this problem as a projection, we can consider a finite sample space $\Omega = \{1, 2, \ldots, n\}$ equipped with the discrete σ-field and the uniform distribution P. The data are represented by two functions $X, Y : \Omega \to \mathbb{R}$, defined by $X(i) = x_i$ and $Y(i) = y_i$, for $i = 1, 2, \ldots, n$. Let I denote the all-one function on Ω.

The mean square error can be expressed as

$$nE[(Y - aX - bI)^2] = n\|Y - aX - bI\|_2^2.$$

Define a subspace of $L^2(P)$ by

$$W = \{aX + bI : a \in \mathbb{R}, \ b \in \mathbb{R}\}.$$

This subspace is closed because it has finite dimension. We can interpret the optimal choice of a and b as the projection of the random variable Y onto the closed subspace W.

12.3   Orthogonality Principle

When W has infinite dimension, computing the projection of a random variable Y onto a closed space W using the definition may be difficult. The next theorem provides a useful characterization of the projection function in terms of perpendicular projection. Essentially, the theorem states that the optimal solution in W that minimizes the distance to Y is achieved when and only when the error vector is perpendicular to W. Note that we are not proving any existence result in the theorem below, as the well-definedness of the projection function has already been established in Theorem 12.4. The objective of the theorem below is to give an alternate way to compute the projection of a random variable.

> **Theorem 12.5 (Orthogonality Principle)**
> *Given a random variable Y in $L^2(P)$ and a closed subspace W in $L^2(P)$, the projection of Y onto W is characterized by*
>
> $$X^* = Proj_W(Y) \iff X^* \in W \text{ and } \langle Y - X^*, X \rangle = 0 \ \forall X \in W.$$

Proof ($\Rightarrow$) Suppose X^* is the random variable in W that minimizes $\|Y - X\|$ over all $X \in W$. We need to show that $\langle Y - X^*, X \rangle = 0$ for all $X \in W$. We proceed by contradiction. Suppose there exists a random variable X_0 in W such that $Y - X^*$ is *not* orthogonal to X_0, i.e., $\langle Y - X^*, X_0 \rangle \neq 0$. We can then deduce that X^* is not the minimizer.

One way to see this by completing the square. Consider

$$\|Y - (X^* - uX_0)\|^2$$

as a function of u. Since X^* minimizes $\|Y - X\|$ over all $X \in W$, this function is minimized at $u = 0$. Expanding it as a polynomial in u, we have

$$\|uX_0 + Y - X^*\|^2 = u^2\|X_0\|^2 + 2u\langle X_0, Y - X^*\rangle + \|Y - X^*\|^2.$$

Since it is assumed that $\langle X_0, Y - X^*\rangle \neq 0$, the L^2 norm $\|X_0\|_2$ should be nonzero. We complete the square and write the above expression as

$$\|X_0\|^2\left(u + \frac{\langle X_0, Y - X^*\rangle}{\|X_0\|^2}\right)^2 - \frac{\langle X_0, Y - X^*\rangle^2}{\|X_0\|^2} + \|Y - X^*\|^2.$$

We can pick the value of u such that the square in the first term is zero. Then,

$$-\frac{\langle X_0, Y - X^*\rangle^2}{\|X_0\|^2} + \|Y - X^*\|^2$$

is strictly less than $\|Y - X^*\|^2$. This contradicts the minimality of $\|Y - X^*\|^2$, and hence, we must have $\langle Y - X^*, X_0 \rangle = 0$.

($\Leftarrow$) Suppose $\hat{X}$ is a random variable such that

$$\langle Y - \hat{X}, X \rangle = 0 \text{ for all } X \in W. \tag{12.2}$$

We want to show that $\hat{X}$ coincides with $\mathrm{Proj}_W(Y)$. We have just proved in the previous paragraph that $\mathrm{Proj}_W(Y) = X^*$ satisfies

$$\langle Y - \mathrm{Proj}_W(Y), X \rangle = 0 \text{ for all } X \in W. \tag{12.3}$$

By subtracting (12.3) from (12.2), we get

$$\langle \mathrm{Proj}_W(Y) - \hat{X}, X \rangle = 0 \text{ for all } X \in W. \tag{12.4}$$

Moreover, $\mathrm{Proj}_W(Y) - \hat{X}$ is a random variable in W because both $\mathrm{Proj}_W(Y)$ and $\hat{X}$ are in W and W is a subspace. Hence, in (12.4), we can substitute X by $\mathrm{Proj}_W(Y) - \hat{X}$ to get $\| \mathrm{Proj}_W(Y) - \hat{X} \|^2 = 0$. This proves that $\mathrm{Proj}_W(Y) = \hat{X}$ almost surely. ∎

In view of the Orthogonality Principle, we can interpret the minimum mean-squared error (MMSE) estimator geometrically as a projection operator.

12.4   Application. MMSE Estimation

The minimum mean-squared error (MMSE) criterion is a widely used approach for optimal linear filtering in signal processing. The Orthogonality Principle provides a powerful tool for deriving the linear MMSE estimator. For more details, readers can refer to textbooks on estimation theory, such as [5]. In this section, we will present some examples of linear and nonlinear MMSE estimation.

12.4.1 Linear MMSE Estimator

Suppose $Y, X_1, X_2, \ldots, X_n$ are in $L^2(P)$. We want to find $a_1, a_2, \ldots, a_n$ such that the mean-squared error

$$\| Y - \sum_{i=1}^{n} a_i X_i \|_2^2$$

is minimized.

Let W be the space consisting of all linear combinations $\sum_{i=1}^{n} a_i X_i$. One can check that this is a closed subspace. By the Orthogonality Principle in Theorem 12.5, we can solve for a_i from

$$\langle Y - \sum_{j=1}^{n} a_j X_j, X_i \rangle = 0, \qquad \text{for } i = 1, 2, \ldots, n.$$

This amounts to solving a system of linear equations:

$$\begin{bmatrix} E[X_1X_1] & E[X_1X_2] & \cdots & E[X_1X_n] \\ E[X_2X_1] & E[X_2X_2] & \cdots & E[X_2X_n] \\ \vdots & \vdots & \ddots & \vdots \\ E[X_nX_1] & E[X_nX_2] & \cdots & E[X_nX_n] \end{bmatrix} \begin{bmatrix} a_1 \\ a_2 \\ \vdots \\ a_n \end{bmatrix} = \begin{bmatrix} E[X_1Y] \\ E[X_2Y] \\ \vdots \\ E[X_nY] \end{bmatrix}.$$

When the matrix on the left is nonsingular, we can solve for a_i's. We remark that when $X_1, \ldots, X_n$ have zero mean, this matrix is the same as the covariance matrix.

Example 12.4.1 (Linear MMSE Estimator with Two Input Random Variables)

When $n = 2$, the problem is to find a_1 and a_2 that minimize $\|Y - a_1X_1 - a_2X_2\|_2$. The optimal solution is given by the following system of linear equations:

$$\begin{bmatrix} E[X_1^2] & E[X_1X_2] \\ E[X_1X_2] & E[X_2^2] \end{bmatrix} \begin{bmatrix} a_1 \\ a_2 \end{bmatrix} = \begin{bmatrix} E[X_1Y] \\ E[X_2Y] \end{bmatrix}.$$

The matrix on the left is positive semi-definite by the Cauchy–Schwarz inequality. The determinant is positive if and only if X_1 and X_2 are linearly independent.

We can use this system of linear equations to solve the linear regression problem in Example 12.2.3, by substituting the expectation $E[X_1^2]$ by $\frac{1}{n}\sum_{i=1}^{n} x_i^2$, the inner product $E[X_1X_2]$ by $\frac{1}{n}\sum_{i=1}^{n} x_i$, etc. By solving

$$\begin{bmatrix} \frac{1}{n}\sum_{i=1}^{n} x_i^2 & \frac{1}{n}\sum_{i=1}^{n} x_i \\ \frac{1}{n}\sum_{i=1}^{n} x_i & 1 \end{bmatrix} \begin{bmatrix} a \\ b \end{bmatrix} = \begin{bmatrix} \frac{1}{n}\sum_{i=1}^{n} x_i y_i \\ \frac{1}{n}\sum_{i=1}^{n} y_i \end{bmatrix},$$

we recover the answer to the least square problem:

$$a = \frac{n\sum_i x_i y_i - \sum_i x_i \sum_i y_i}{n\sum_i x^2 - (\sum_i x_i)^2},$$

$$b = \frac{\sum_i y_i - a\sum_i x_i}{n}.$$

Example 12.4.2 (Scaling Factor for Linear MMSE Estimate)

Suppose we want to estimate a random variable X, which is observed with some additive noise N. The observed random variable is $Y = cX + N$, where c is a nonzero constant. We suppose that X has mean μ and variance σ_X^2, N has zero mean and variance σ_N^2, and X and N are independent. We also suppose that the parameters c, μ, σ_X^2, and σ_N^2 are known, and $c \neq 0$.

An unbiased estimate of X is simply $X = Y/c$, if we can measure the random variable Y. However, if we want to minimize the mean-squared error, the optimal constant k that minimizes the mean-squared error $\|kY - X\|^2$ can be computed by differentiating

$$k^2 E[Y^2] - 2kE[XY] + E[X^2]$$

with respect to k. The optimal value scaling coefficient is

$$k^* = \frac{E[XY]}{E[Y^2]} = \frac{c\sigma_X^2}{c^2\sigma_X^2 + \sigma_N^2}.$$

The linear MMSE estimate of X is then given by

$$\hat{X}_{\text{MMSE}} = \frac{c^2 \sigma_X^2}{c^2 \sigma_X^2 + \sigma_N^2} \cdot \frac{Y}{c}.$$

We multiply the unbiased estimate Y/c by the scaling factor $\frac{c^2 \sigma_X^2}{c^2 \sigma_X^2 + \sigma_N^2}$ to obtain the MMSE estimate, which is a linear function of the observed variable Y.

12.4.2 Nonlinear MMSE Estimation

In general, given $Y, X_1, X_2, \ldots, X_n \in L^2(P)$, the problem of nonlinear MMSE estimation is to find a measurable function $g(x_1, x_2, \ldots, x_n)$ such that

$$\|Y - g(X_1, X_2, \ldots, X_n)\|_2$$

is minimized. We formulate the problem in terms of σ-algebra.

Recall that $\sigma(X_1, X_2, \ldots, X_n)$ denotes the smallest σ-algebra of $\mathscr{F}$ such that $X_1, \ldots, X_n$ are measurable. A function $g(x_1, x_2, \ldots, x_n)$ mapping from $\mathbb{R}^n$ to $\mathbb{R}$ is measurable with respect to $\sigma(X_1, X_2, \ldots, X_n)$ iff

$$g^{-1}(B) \in \sigma(X_1, X_2, \ldots, X_n) \text{ for all } B \in \mathscr{B}(\mathbb{R}).$$

The next theorem characterizes the random variables that can be written as a measurable function of X_1 to X_n.

Theorem 12.6 (Functions Measurable with Respect to $\sigma(\mathbf{X})$)

$$\mathbf{X}(\omega) = (X_1(\omega), X_2(\omega), \ldots, X_n(\omega))$$

Consider an $\mathscr{F}$-measurable vector function $\mathbf{X}(\omega) = (X_1(\omega), X_2(\omega), \ldots, X_n(\omega))$ defined on a measurable space $(\Omega, \mathscr{F})$. An $\mathscr{F}$-measurable function $Y(\omega)$ can be written as $Y = g(X_1, X_2, \ldots, X_n)$ for a $(\mathscr{B}(\mathbb{R}^n), \mathscr{B}(\mathbb{R}))$-measurable function g if and only if Y is measurable with respect to $\sigma(\mathbf{X})$.

Proof ($\Rightarrow$) Suppose that $\mathbf{X}$ is $\mathscr{F}$-measurable from Ω to $\mathbb{R}^n$, and g is $(\mathscr{B}(\mathbb{R}^n), \mathscr{B}(\mathbb{R}))$-measurable function from $\mathbb{R}^n$ to $\mathbb{R}$. Consider the composite function $Y = g \circ \mathbf{X}$. Given any Borel set $B \in \mathscr{B}(\mathbb{R})$, the inverse image $g^{-1}(B)$ is $\mathscr{B}(\mathbb{R}^n)$-measurable. Since the vector function $\mathbf{X}$ is $(\sigma(\mathbf{X}), \mathscr{B}(\mathbb{R}^n))$-measurable, the inverse image $Y^{-1}(B) = \mathbf{X}^{-1}(g^{-1}(B))$ is in $\sigma(\mathbf{X})$.

($\Leftarrow$) Suppose that Y is an $\mathscr{F}$-measurable function that is also $\sigma(\mathbf{X})$-measurable. By definition, we have $Y^{-1}(B) \in \sigma(\mathbf{X})$ for all Borel sets B in $\mathbb{R}$.

When Y is an indicator function $\mathbf{1}_A(\omega)$ for some $A \in \mathscr{F}$, the inverse image $Y^{-1}(\{1\}) = A$ must be in $\sigma(\mathbf{X})$ by the assumption that Y is $\sigma(\mathbf{X})$-measurable. Hence, $A = \mathbf{X}^{-1}(B)$ for some $B \in \mathscr{B}(\mathbb{R}^n)$. We can set

$$g(x_1, x_2, \ldots, x_n) = \begin{cases} 1 & \text{if } (x_1, x_2, \ldots, x_n) \in B, \\ 0 & \text{otherwise.} \end{cases}$$

Then g is Borel measurable, and $Y(\omega) = g(\mathbf{X}(\omega))$ for all $\omega \in \Omega$.

Next, suppose Y is a simple function in the form $\sum_{i=1}^m a_i \mathbf{1}_{A_i}$, for some $A_i \in \mathscr{F}$. We can find Borel measurable function g_i such that

$$Y(\omega) = \sum_{i=1}^m a_i g_i(X_1(\omega), \ldots, X_n(\omega)).$$

The rest of the proof is to extend it to nonnegative measurable functions and real-valued measurable functions. It is the standard machinery of Lebesgue integral and the detail is omitted. ∎

In nonlinear MMSE estimation, we define W to be the set of all random variables on $(\Omega, \mathscr{F}, P)$ that are measurable with respect to the σ-algebra $\sigma(\mathbf{X})$ and have finite second moment. We denote this set by $L^2(\Omega, \sigma(\mathbf{X}), P)$, where P denotes the restriction of the original probability measure to the sub-σ-algebra $\sigma(\mathbf{X})$. Note that $L^2(\Omega, \sigma(\mathbf{X}), P)$ is a closed subspace of the space of all square-integrable random variables on $(\Omega, \mathscr{F}, P)$.

The optimization problem can now be stated as

$$\min_{g(x_1, \ldots, x_n)} \|Y - g(X_1, X_2, \ldots, X_n)\|_2$$

with the minimum taken over all functions $g : \mathbb{R}^n \to \mathbb{R}$ such that $g(X_1, \ldots, X_n) \in L^2(\Omega, \sigma(\mathbf{X}), P)$. We project Y onto the closed subspace $L^2(\Omega, \sigma(\mathbf{X}), P)$. By Theorem 12.4, a solution exists and any two solutions are equal almost surely. By the Orthogonal Principle, the solution is characterized by

$$E[\hat{Y}Z] = E[YZ] \qquad \text{for all } Z \in L^2(\Omega, \sigma(\mathbf{X}), P). \tag{12.5}$$

Indeed, the condition in (12.5) can be written in several equivalent forms,

$$\langle Y - \hat{Y}, Z \rangle = 0 \qquad \text{for all } Z \in L^2(\Omega, \sigma(\mathbf{X}), P)$$

$$\Leftrightarrow E[(Y - \hat{Y})Z] = 0 \qquad \text{for all } Z \in L^2(\Omega, \sigma(\mathbf{X}), P)$$

$$\Leftrightarrow E[YZ] = E[\hat{Y}Z] \qquad \text{for all } Z \in L^2(\Omega, \sigma(\mathbf{X}), P)$$

$\Leftrightarrow E[Yh(\mathbf{X})] = E[\hat{Y}h(\mathbf{X})]$ for all $\mathscr{B}(\mathbb{R}^n)$-measurable $h(\mathbf{x})$ s.t. $h(\mathbf{X}) \in L^2(P)$.

We illustrate this idea in the following example.

Example 12.4.3 (An Example of Nonlinear MMSE Estimation)
Consider a sample space $\{a, b, c, d\}$ and define a uniform distribution on the four sample points. Define two random variables using the table

ω	a	b	c	d
$X(\omega)$	1	1	2	2
$Y(\omega)$	1	2	3	4

We want to find the nonlinear MMSE estimator of Y given X. Since the probability space is finite, all functions defined on this probability space are measurable and have finite second moments. The problem is equivalent to finding a function $g(x)$ that minimizes $\|Y - g(X)\|_2$.
The σ-algebra generated by X contains four sets $\emptyset$, Ω, $\{a, b\}$, and $\{c, d\}$. The MMSE estimator of Y given X is a function of X and is in the form

$$\hat{Y}(\omega) = \begin{cases} \alpha & \text{if } \omega = a \text{ or } \omega = b \\ \beta & \text{if } \omega = c \text{ or } \omega = d. \end{cases}$$

The optimal values of α and β can be determined using calculus. The objective function that we are going to minimize is a multi-variable quadratic function,

$$\frac{1}{4}(1 - \alpha)^2 + \frac{1}{4}(2 - \alpha)^2 + \frac{1}{4}(3 - \beta)^2 + \frac{1}{4}(4 - \beta)^2.$$

The solution can be derived by partially differentiating the objective function with respect to α and β and solving a system of linear equations.
In the followings, we can obtain the answer using the Orthogonal Principle. The MMSE estimator $\hat{Y}$ satisfies the following property:

$$E[YZ] = E[\hat{Y}Z] \quad \text{for all } \sigma(X)\text{-measurable functions } Z.$$

We first consider $Z = \mathbf{1}_{\{a,b\}}$. Applying (12.5), we get

$$E[Y\mathbf{1}_{\{a,b\}}] = \frac{1}{4} + \frac{2}{4} + 0 + 0,$$

$$E[\hat{Y}\mathbf{1}_{\{a,b\}}] = \alpha E[\mathbf{1}_{\{a,b\}}] = \alpha\left(\frac{1}{4} + \frac{1}{4}\right).$$

The second line follows from the fact that $\hat{Y}$ is constant on the event $\{a, b\}$. Equating the above two values yields $\alpha = 1.5$.
We next consider the function $Z = \mathbf{1}_{\{c,d\}}$. We have

$$E[Y\mathbf{1}_{\{c,d\}}] = 0 + 0 + \frac{3}{4} + \frac{4}{4},$$

$$E[\hat{Y}\mathbf{1}_{\{c,d\}}] = \beta E[\mathbf{1}_{\{c,d\}}] = \beta\left(\frac{1}{4} + \frac{1}{4}\right).$$

This gives $\beta = 3.5$.

The optimal function $g(x)$ that minimizes mean-squared error $\|Y - g(X)\|_2$ is

$$g(x) = \begin{cases} 1.5 & \text{if } x = 1, \\ 3.5 & \text{if } x = 2. \end{cases}$$

We implement this example using Python.

```
from random import choice

Omega = ['a','b','c','d']          # sample space
Y = {'a':1, 'b':2, 'c':3, 'd':4}   # represent Y by a dictionary

omega= choice(Omega)               # pick a sample uniformly
X = 1 if (Y[omega] <= 2) else 2    # X equals 1 if Y=1, 2 if Y=2
Y_hat = 1.5 if X==1 else 3.5       # estimate random variable Y by X

print(f"Y={Y[omega]}, X={X}, Y_hat={Y_hat}")
```

In the program, we denote the estimate of Y by Y_hat, which is a function of X. A sample output of this program is:

```
Y=4, X=2, Y_hat=3.5
```

The conditional expectation operator in $L^2(P)$ generally provides the nonlinear minimum mean-squared error estimation. Specifically, the function $g(x)$ that minimizes the mean-squared error $E[(Y - g(X))^2]$ is given by the conditional expectation $g(x) = E[Y|X = x]$.

Problems

12.1. Derive the parallelogram law $\|\mathbf{u} + \mathbf{v}\|_2^2 + \|\mathbf{u} - \mathbf{v}\|_2^2 = 2\|\mathbf{u}\|_2^2 + 2\|\mathbf{v}\|_2^2$, which holds for any vectors $\mathbf{u}$ and $\mathbf{v}$ in a Hilbert space with L^2 norm $\|\cdot\|_2$.

12.2. Consider an inner product space with induced L^2 norm denoted by $\|\cdot\|_2$.

(a) Prove the continuity of the L^2 norm, that is, if $(X_k)_{k\geq 1}$ is a sequence in the inner product space such that $\|X_k - X\|_2 \to 0$ for some X as $n \to \infty$, then $\lim_{n\to\infty} \|X_k\|_2 = \|X\|_2$.

(b) Let $(X_k)_{k\geq 1}$ be a Cauchy sequence. Show that the L^2 norms of the sequence are bounded, i.e., there exists a real number M such that $\|X_k\|_2 \leq M$ for all k.

(c) Suppose the Cauchy $(X_k)_{k\geq 1}$ in part (b) converges to X in L^2 norm. Prove that $\|X\|_2$ is finite.

12.3. Denote the pdf of the exponential distribution with mean 1 by

$$f(x) = \begin{cases} e^{-x} & \text{if } x > 0 \\ 0 & \text{otherwise.} \end{cases}$$

Define an inner product $\langle g, h \rangle \triangleq \int_0^\infty g(x)h(x)f(x)\,dx$ for two functions $g(x)$ and $h(x)$. Compute the projection of the square function $g(x) = x^2$ onto the subspace that consists of constant functions.

12.4. Consider a finite sample space $\Omega = \{a, b, c, d, e\}$ with a uniform probability measure. Define two random variables X and Y by

ω	a	b	c	d	e
$X(\omega)$	1	2	1	2	3
$Y(\omega)$	1	2	3	4	5

(a) Derive the MMSE estimator of Y as a function of X.
(b) Derive the MMSE estimator of X as a function of Y.

12.5. We model the temperature in a forest by a random variable X. Suppose that the mean μ and the variance σ_0^2 are known. The temperature is measured by a sensor. The output of the sensor $Y_1 = X + N_1$ is corrupted by Gaussian noise $N_1 \sim N(0, \sigma_1^2)$. Assume that N_1 and X are independent, and the variance σ_1^2 is known.

(a) We can estimate X by $\alpha(Y_1 - \mu) + \mu$. Find the optimal choice of constant α that minimizes the mean-squared error $E[(X - \mu - \alpha(Y_1 - \mu))^2]$.
(b) We install another temperature sensor. The measurement of the second sensor is $Y_2 = X + N_2$, where $N_2 \sim N(0, \sigma_2^2)$. Assume X, N_1, and N_2 are independent, and the variance σ_2^2 is known. Find the optimal values of α and β that minimize

$$E[(X - \mu - \alpha(Y_1 - \mu) - \beta(Y_2 - \mu))^2].$$

Conditional Expectation

13

As a motivating example, suppose we want to obtain the average height of people in a country by taking uniform samples from the population. We define the set of people in this country as Ω, and the uniform distribution on Ω as P. We denote the height of a person ω chosen from the population by $X(\omega)$. The expectation $E[X]$ represents the average height of the population. However, we can also compute the average height in the country by first determining the average height in each province. Let $E[X|$ province $A]$ denote the average height in a given province A. The national average height can then be computed as a weighted average of the provincial averages, with the weighting factors proportional to the population in the provinces.

We can also partition the people in this country by the cities they live in. The average height in a city is the conditional expectation $E[X|$ city$]$. The partition of the country into cities is a more refined partition compared to the partition into provinces. This example illustrates that the conditional expectation is closely related to how we partition the sample space.

In this chapter, we first review the conditional expectation of a random variable given an event. Then we define the conditional expectation conditioned on a σ-algebra generated by a finite number of events. In the abstract definition of conditional expectation, we will condition on a general σ-algebra. Finally, we will discuss filtrations and martingales as applications of conditional expectation.

Throughout this chapter, we will work with a fixed probability space $(\Omega, \mathcal{F}, P)$.

13.1 Expectation Conditioned on a Finite Partition

In probability theory, we define the conditional probability of an event E given another event A by $P(A \cap E)/P(A)$, provided that $P(A) > 0$. This quantity represents the likelihood of the event E given that we already know that the event A has occurred.

© The Author(s), under exclusive license to Springer Nature Switzerland AG 2023   211
K. Shum, *Measure-Theoretic Probability*, Compact Textbooks in Mathematics,
https://doi.org/10.1007/978-3-031-49830-5_13

If the event A is fixed and the event E varies, we obtain a measure function whose value is conditional probability.

Definition 13.1

Let A be an event in $\mathcal{F}$ with $P(A) \neq 0$. We define a probability measure μ_A by

$$\mu_A(E) \triangleq \frac{P(A \cap E)}{P(A)}.$$

The *expectation conditioned on the event* A is defined by the integral with respect to the measure μ_A,

$$E[X|A] \triangleq \int_\Omega X \, d\mu_A.$$

It is easy to verify that μ_A is a probability measure on $\mathcal{F}$ with $\mu_A(A) = 1$. In the calculations of $E[X]$ and $E[X|A]$, we have the same function X mapping from Ω to the real numbers. The difference is in the change of measure from P to μ_A.

The next theorem gives the relationship between the two expectation operators.

Theorem 13.1
For a random variable X in $L^1(P)$,

$$E[X|A] = \frac{1}{P(A)} \int_A X \, dP.$$

Proof Suppose X is an indicator function $\mathbf{1}_B$ for some $B \in \mathcal{F}$. Then we have

$$E[\mathbf{1}_B|A] \triangleq \int_\Omega \mathbf{1}_B \, d\mu_A = \mu_A(B) \triangleq \frac{P(A \cap B)}{P(A)}.$$

On the other hand,

$$\frac{1}{P(A)} \int_A \mathbf{1}_B \, dP = \frac{1}{P(A)} \int_\Omega \mathbf{1}_A \mathbf{1}_B \, dP = \frac{1}{P(A)} \int_\Omega \mathbf{1}_{A \cap B} \, dP = \frac{P(A \cap B)}{P(A)}.$$

Therefore, the equality in the theorem holds for $\mathcal{F}$-measurable indicator functions. By linearity, we can extend it to all $\mathcal{F}$-measurable simple functions.

The remaining tasks are to extend this to nonnegative functions and real-valued functions. By the monotone convergence theorem, we can extend it to all nonnegative $\mathcal{F}$-measurable functions. Finally, by decomposing real-valued functions into positive and negative parts, we can extend it to all $\mathscr{F}$-measurable real-valued functions. The rest of the proof is mechanical and is omitted. ∎

We next extend the definition of conditional expectation to conditional expectation given a finite partition, as we would like to have the flexibility to condition on different sets A.

Definition 13.2

Consider a finite partition $A_1, A_2, \ldots, A_k$ of the sample space Ω. Let $\mathscr{G}$ be the σ-algebra generated by $A_1, A_2, \ldots, A_k$. For any $\mathscr{G}$-measurable function in $L^1(P)$, define the *conditional expectation given $\mathscr{G}$* by

$$E[X|\mathscr{G}] \triangleq \sum_{i=1}^{k} E[X|A_i] \mathbf{1}_{A_i}.$$

A few important notes are in order:

- The conditional expectation given $\mathscr{G}$ is a random variable. We can write it as $E[X|\mathscr{G}](\omega)$ to emphasize that the input of the function is ω.
- $E[X|\mathscr{G}]$ is a simple function that is $\mathscr{G}$-measurable and constant on each A_i. The value of $E[X|\mathscr{G}](\omega)$ for $\omega \in A_i$ is $E[X|A_i]$.
- If $A_i = \emptyset$ or $P(A_i) = 0$, the conditional expectation $E[X|A_i]$ given A_i is not defined, but $E[X|\mathscr{G}]$ is still well-defined.

Since $E[X|\mathscr{G}]$ is a random variable, a natural question is how to compute the expectation.

Theorem 13.2
For any event $C \in \mathscr{G}$,

$$\int_C E[X|\mathscr{G}] \, dP = \int_C X \, dP. \tag{13.1}$$

In particular, we have

$$E[E[X|\mathscr{G}]] = E[X]. \tag{13.2}$$

Proof Recall that the events A_1 to A_k form a partition of Ω and are mutually disjoint. They are the atoms of the σ-algebra $\mathscr{G}$. Any event C in $\mathscr{G}$ is a union of some sets among A_1 to A_k. Thus, we can find an index set $I \subseteq \{1, 2, \ldots, k\}$ such that $C = \cup_{i \in I} A_i$.

By the definition of expectation for simple function, we have

$$\int_C E[X|\mathcal{G}] \, dP = \sum_{i=1}^{k} E[X|A_i] \cdot P(C \cap A_i) = \sum_{i \in I} E[X|A_i] P(A_i)$$

$$= \sum_{i \in I} \left(\frac{1}{P(A_i)} \int_{A_i} X \, dP \right) P(A_i),$$

which can be simplified to

$$\sum_{i \in I} \int_{A_i} X \, dP = \int_C X \, dP.$$

This holds for any $C \in \mathcal{G}$. By setting $C = \Omega$, we obtain (13.2).   ■

The equality in (13.2) is saying that the overall average value $E[X]$ is a weighted average of the local average values $E[X|A_i]$'s. In other words, the value of $E[X|\mathcal{G}]$ can be thought of as a "smoothing" of the random variable X that takes the partition $A_1, A_2, \ldots, A_k$ into account.

13.2   Expectation Conditioned on a Sub-sigma-algebra

We extend the notion of conditional expectation to the general case by replacing the special σ-algebra $\mathcal{G}$ in Definition 13.2 by any sub-σ-algebra of $\mathcal{F}$.

Definition 13.3

Let X be a real-valued P-integrable random variable and $\mathcal{G}$ is a sub-σ-algebra of $\mathcal{F}$. The *conditional expectation of X given $\mathcal{G}$*, denoted by $E[X|\mathcal{G}]$, is a $\mathcal{G}$-measurable random variable that satisfies

$$\int_B E[X|\mathcal{G}] \, dP = \int_B X \, dP \qquad \text{for all } B \in \mathcal{G}. \qquad (13.3)$$

The condition in (13.3) can be written in two equivalent forms. These forms provide alternative ways to think about the definition of $E[X|\mathcal{G}]$.

(i)

$$\int_\Omega E[X|\mathcal{G}] \cdot \mathbf{1}_B \, dP = \int_\Omega X \cdot \mathbf{1}_B \, dP \qquad \text{for all } B \in \mathcal{G}; \qquad (13.4)$$

(ii)

$$\int_\Omega E[X|\mathcal{G}] \cdot Z \, dP = \int_\Omega X \cdot Z \, dP \qquad (13.5)$$

for all bounded $\mathcal{G}$-measurable random variables Z.

It is obvious that condition (ii) in (13.5) implies condition (i) in (13.4) because the indicator function $\mathbf{1}_B$ for any $B \in \mathcal{G}$ is $\mathcal{G}$-measurable. To show the reverse direction, we can proceed with the sequence of indicator functions, simple functions, and bounded functions that are $\mathcal{G}$-measurable.

Comparing condition (ii) in (13.5) with the version of Orthogonal Principle in (12.5), we see that when X is square integrable, the conditional expectation of X given a sub-σ-algebra $\mathcal{G}$ is the same as the projection of X on the space of $\mathcal{G}$-measurable functions with finite second moments. Roughly speaking, for random variables in $L^2(P)$, the conditional expectation is just a minimum mean-squared error (MMSE) estimator. A more precise statement is given in the next theorem.

Theorem 13.3

If X is in $L^2(P)$, then $E[X|\mathcal{G}]$ is the MMSE estimator of X that achieves the minimum

$$\min_Z \|Z - X\| = \|E[X|\mathcal{G}] - X\|$$

with the minimum taken over all $\mathcal{G}$-measurable random variables in $L^2(P)$.

The example of nonlinear MMSE estimator in Example 12.4.3 is indeed an example of conditional expectation in the case of a finite sample space.

▶ **Remark 13.1** When $\mathcal{G}$ is generated by a partition of Ω consisting of k disjoint events, then $E[X|\mathcal{G}]$ reduces to the conditional expectation defined in the previous section.

We first prove that the condition in (13.3) uniquely defines the conditional expectation, up to a null set. That is, if there are two solutions to (13.3), then they must be equal almost surely.

Theorem 13.4 (Uniqueness of Conditional Expectation)

If Z_1 and Z_2 are $\mathcal{G}$-measurable random variables that satisfy

---

(continued)

Theorem 13.4 (continued)

$$\int_B Z_1 \, dP = \int_B Z_2 \, dP$$

for all $B \in \mathcal{G}$, then $Z_1 = Z_2$ a.s.

Proof Let A be the event $\{\omega : Z_1(\omega) \geq Z_2(\omega)\}$. The event A is $\mathcal{G}$-measurable because $Y = Z_1 - Z_2$ is $\mathcal{G}$-measurable and $A = Y^{-1}([0, \infty))$.

Because A is in $\mathcal{G}$ and the expectation operator is linear, we have

$$0 = \int_A Z_1 - Z_2 \, dP.$$

However, $Z_1(\omega) - Z_2(\omega) \geq 0$ for ω in A. The above equation can hold only when $Z_1(\omega) = Z_2(\omega)$ almost everywhere on A, meaning that there is an event $E_1 \subset A$ with $P(E_1) = 0$ such that $Z_1(\omega) = Z_2(\omega)$ for all $\omega \in A \setminus E_1$.

Likewise, consider the event $B = \{\omega : Z_1(\omega) \leq Z_2(\omega)\}$. We get $Z_1(\omega) = Z_2(\omega)$ almost everywhere on B. There is an event $E_2 \subset B$ with $P(E_2) = 0$ and $Z_1(\omega) = Z_2(\omega)$ for all $\omega \in B \setminus E_2$.

We complete the proof by noting that $A \cup B = \Omega$ and $P(E_1 \cup E_2) = 0$. ∎

Theorem 13.4 says that if Y is a random variable that satisfies (13.3), any other random variable Y' that satisfies (13.3) is equal to Y with probability 1. On the other hand, if Y is a random variable that satisfies (13.3), then any random variable that equals Y a.s. is also a solution to (13.3). A solution to (13.3) is called a *version* of $E[X|\mathcal{G}]$.

We illustrate the abstract definition of conditional expectation by considering the smallest and the largest possible sub-σ-algebras.

Example 13.2.1 (Conditioning on the Smallest σ-Algebra)
Take $\mathcal{G} = \{\emptyset, \Omega\}$. A $\mathcal{G}$-measurable random variable X is a constant function. That is, there is a constant c such that $X(\omega) = c$ for all $\omega \in \Omega$. We want to determine the value of c. In (13.3), if we take $B = \Omega$, we get

$$\int_\Omega E[X|\{\emptyset, \Omega\}] \, dP = E[X].$$

On the other hand, since $E[X|\mathcal{G}](\omega)$ is equal to the constant c for all ω, we have

$$\int_\Omega E[X|\{\emptyset, \Omega\}] \, dP = \int_\Omega c \, dP = c.$$

Therefore, the value of c is equal to $E[X]$. The conditional expectation $E[X|\{\emptyset, \Omega\}]$ is a constant function with value $E[X]$.

Example 13.2.2 (Conditioning on the Largest σ-Algebra)

Consider the case $\mathscr{G} = \mathscr{F}$. Let B be any set in $\mathscr{G}$, i.e., B is any set in $\mathscr{F}$. We want to find a $\mathscr{G}$-measurable function, which is the same as an $\mathscr{F}$-measurable function $E[X|\mathscr{F}]$ such that (13.3) holds for all $B \in \mathscr{F}$. In fact, X is such a choice for $E[X|\mathscr{F}]$. Hence, $E[X|\mathscr{F}] = X$ a.s.

We define $E[X|Y]$ in terms of the σ-algebra generated by Y.

Definition 13.4

The *conditional expectation of a random variable X given random variable Y* is defined as

$$E[X|Y] \triangleq E[X|\sigma(Y)].$$

For a family of random variables $\{Y_i\}_{i \in I}$, the conditional expectation of X given $\{Y_i\}_{i \in I}$ is defined as

$$E[X|\{Y_i\}_{i \in I}] \triangleq E[X|\sigma(\{Y_i\}_{i \in I})].$$

In particular, the conditional expectation of X given random variables Y and Z is defined as

$$E[X|Y, Z] \triangleq E[X|\sigma(Y, Z)].$$

Conditional probability is defined in terms of conditional expectation.

In the above definition, $\sigma(Y, Z)$ means the smallest σ-algebra under which both Y and Z are measurable, and $\sigma(\{Y_i\}_{i \in I})$ denotes the smallest σ-algebra such that all random variables Y_i's are measurable.

Definition 13.5

Let $(\Omega, \mathscr{F}, P)$ be a probability space. The *conditional probability* of an event A in $\mathscr{F}$ given a random variable X is defined as

$$P(A|X) \triangleq E[\mathbf{1}_A|\sigma(X)].$$

When the conditional expectation $E[X|Y]$ as regarded as a function of Y, we write it as $E[X|Y = y]$. This is a function from $\mathbb{R}$ to $\mathbb{R}$, with y as the argument. However, the more fundamental definition of conditional expectation $E[X|Y]$ is a function from Ω to $\mathbb{R}$, which is measurable with respect to σ-field $\sigma(Y)$.

13.3 Properties of Conditional Expectation

In this section, we consider a probability space $(\Omega, \mathscr{F}, P)$ and a sub-σ-field $\mathscr{G}$ of $\mathscr{F}$. We first address the existence of conditional expectation, which follows from the Radon–Nikodym theorem.

Let μ and ν be measures defined on the same measurable space. We say that measure ν is *absolutely continuous* with respect to measure μ if $\mu(A) = 0 \Rightarrow \nu(A) = 0$. This is denoted by $\nu \ll \mu$.

For example, a continuous-type distribution is absolutely continuous with respect to the Lebesgue measure. The theorem by Radon and Nikodym provides a partial converse result. We state it below without proof.

Theorem 13.5 (Radon–Nikodym)

Let $\mathcal{G}$ be a sub-σ-field of $\mathcal{F}$. Suppose ν and μ are σ-finite measures such that $\nu \ll \mu$. Then there exists a nonnegative $\mathcal{G}$-measurable function f such that $\nu(B) = \int_B f \, d\mu$ for all $B \in \mathcal{G}$. The function f is unique in the sense that if g is another $\mathcal{G}$-measurable function with the same property, then g is equal to f μ-almost everywhere.

The function f is called the density *or the* Radon–Nikodym derivative *of ν with respect to μ. It is represented by the symbol $\frac{d\nu}{d\mu}$.*

Theorem 13.6 (Existence of Conditional Expectation)

If $X \in L^1(P)$ and $\mathcal{G}$ is a sub-σ-field of $\mathcal{F}$, then the conditional expectation $E[X|\mathcal{G}]$ exists.

Proof For random variable X in $L^2(P)$, we know that the conditional expectation of X can be obtained by the projection function.

For random variable X in $L^1(P)$, the existence of conditional expectation is a consequence of the Radon–Nikodym theorem because the measure ν defined by

$$\nu(B) = \int_B X \, dP$$

is absolutely continuous with respect to measure P. We can take the Radon–Nikodym derivative $d\nu/dP$ as the conditional expectation.   ∎

Most of the random variables we encounter have finite second moment. The construction of conditional expectation using L^2 theory and projection suffices for most purposes. However, when a random variable is in L^1 but not in L^2, we need to use the Radon–Nikodym theorem to prove the existence. We refer readers to more advanced textbooks, such as [2] and [4], for further details.

From now on, we will assume that conditional expectation for random variable in L^1 exists.

Theorem 13.7 (Properties of Conditional Expectation)
Let $(\Omega, \mathcal{F}, P)$ denote a probability space, and let $\mathcal{G}$ be a σ-subfield of $\mathcal{F}$:

1. *If X is in $L^1(P)$ and is $\mathcal{G}$-measurable, then $E[X|\mathcal{G}] = X$ a.s. In particular, for any constant c, $E[c|\mathcal{G}] = c$ a.s.*
2. *(Nonnegativity) If $X \geq 0$ a.s., then $E[X|\mathcal{G}] \geq 0$ a.s.*
3. *(Linearity) Suppose $X, Y \in L^1(P)$ and a is a constant. We have*

$$E[X + Y|\mathcal{G}] = E[X|\mathcal{G}] + E[Y|\mathcal{G}],$$

and

$$E[aX|\mathcal{G}] = aE[X|\mathcal{G}].$$

4. *(Monotonicity) If X and Y are in $L^1(P)$ and $X \leq Y$ a.s., then $E[X|\mathcal{G}] \leq E[Y|\mathcal{G}]$ a.s.*
5. *(Conditional triangle inequality) For random variable X in $L^1(P)$,*

$$\left| E[X|\mathcal{G}] \right| \leq E\big[|X|\,|\mathcal{G}\big].$$

Proof

(1) We verify that X is $\mathcal{G}$-measurable, and we can replace $\hat{X}$ by X in the defining property $E[\hat{X}\mathbf{1}_B] = E[X\mathbf{1}_B]$ for all $B \in \mathcal{G}$. The second statement in (1) follows because a constant random variable is $\mathcal{G}$-measurable for any σ-algebra $\mathcal{G}$.

(2) Suppose $X \geq 0$ almost surely, and $\hat{X}$ is a version of $E[X|\mathcal{G}]$. For integer $n \geq 1$, take E_n to be the set $\{\omega : \hat{X} \leq -1/n\}$, for integer $n \geq 1$. The set E_n is $\mathcal{G}$-measurable because $\hat{X}$ is $\mathcal{G}$-measurable. Because X is nonnegative a.s. by assumption,

$$\int_{E_n} \hat{X} = \int_{E_n} X \geq 0$$

for all n. Hence,

$$0 \leq \int_{E_n} \hat{X}\,dP \leq \int_{E_n} \left(-\frac{1}{n}\right) dP = -\frac{1}{n} P(E_n).$$

This is possible only if $P(E_n) = 0$. Since $P(E_n) = 0$ for all $n \geq 1$, the event $\{\hat{X} < 0\} = \cup_n E_n$ must have probability zero. This proves that $\hat{X} \geq 0$ a.s.

(3) For any event $B \in \mathscr{G}$, we have

$$\int_B X = \int_B \hat{X} \tag{13.6}$$

$$\int_B Y = \int_B \hat{Y}, \tag{13.7}$$

where $\hat{X}$ and $\hat{Y}$ are versions of $E[X|\mathscr{G}]$ and $E[Y|\mathscr{G}]$, respectively. By adding (13.6) and (13.7), we obtain

$$\int_B Z = \int_B (X + Y) = \int_B (\hat{X} + \hat{Y}). \tag{13.8}$$

Comparing with the defining property of conditional expectation for $E[Z|\mathscr{G}]$,

$$\int_B Z = \int_B \hat{Z},$$

where $\hat{Z}$ denotes a version of $E[Z|\mathscr{G}]$, we see that

$$\int_B \hat{Z} = \int_B (\hat{X} + \hat{Y})$$

for all $B \in \mathscr{G}$. This proves $\hat{Z} = \hat{X} + \hat{Y}$ a.s.

Similarly, we can prove $E[aX|\mathscr{G}] = a\hat{X}$ a.s.

(4) Since $Y - X \geq 0$ a.s., by the nonnegativity of conditional expectation, we have $E[Y - X|\mathscr{G}] \geq 0$ a.s. By linearity, we obtain $E[Y|\mathscr{G}] - E[X|\mathscr{G}] \geq 0$.

(5) Since $X \leq |X|$, by monotonic property of conditional expectation,

$$E[X|\mathscr{G}] \leq E[|X||\mathscr{G}] \quad a.s.$$

Similarly, from $-X \leq |X|$, we derive that

$$-E[X|\mathscr{G}] = E[-X|\mathscr{G}] \leq E[|X||\mathscr{G}] \quad a.s.$$

This proves that the absolute value of the conditional expectation of X given $\mathscr{G}$ is less than or equal to the conditional expectation of $|X|$ given $\mathscr{G}$.

■

The properties listed in the previous theorem all have analog in ordinary expectation. The basic convergence theorems such as Fatou's lemma, monotone convergence theorem, and dominated convergence theorem also have versions for conditional expectation. However, the next three properties are unique to conditional expectation.

Theorem 13.8 (Tower Property)
Suppose $\mathcal{F} \supseteq \mathcal{G} \supseteq \mathcal{H}$ is a tower of σ-algebras and X is in $L^1(P)$. Then

$$E[X|\mathcal{H}] = E[\,E[X|\mathcal{G}]|\mathcal{H}\,] \quad a.s.$$

Proof We want to show that for all $B \in \mathcal{H}$,

$$\int_B E[X|\mathcal{H}]\,dP = \int_B E[\,E[X|\mathcal{G}]|\mathcal{H}\,]\,dP. \tag{13.9}$$

The left-hand side is $\int_B X\,dP$. The right-hand side is equal to $\int_B E[X|\mathcal{G}]\,dP = \int_B X\,dP$. The first equality is due to $B \in \mathcal{H}$. The second equality follows from $B \in \mathcal{G}$. Therefore, both sides of (13.9) are identical.

Since (13.9) holds for all $\mathcal{H}$-measurable set B, we conclude that $E[\,E[X|\mathcal{G}]|\mathcal{H}\,]$ is a version of the conditional expectation of X given $\mathcal{H}$. ∎

Theorem 13.9
Suppose XY are in $L^1(P, \mathcal{F})$, where X is a $\mathcal{G}$-measurable function and Y is $\mathcal{F}$-measurable and integrable, then

$$E[XY|\mathcal{G}] = X E[Y|\mathcal{G}] \quad a.s.$$

In particular, when X is a function of Z, we have $E[XY|Z] = X E[Y|Z]$ a.s.

Proof We first show that, for any $\mathcal{G}$-measurable set A,

$$\int_B E[\mathbf{1}_A Y|\mathcal{G}]\,dP = \int_B \mathbf{1}_A E[Y|\mathcal{G}]\,dP \quad \forall B \in \mathcal{G}. \tag{13.10}$$

For each $B \in \mathcal{G}$, the right-hand side of (13.10) equals

$$\int \mathbf{1}_{B \cap A} E[Y|\mathcal{G}]\,dP = \int_{B \cap A} E[Y|\mathcal{G}]\,dP = \int_{B \cap A} Y\,dP. \quad (\because B \cap A \in \mathcal{G}).$$

Meanwhile, the left-hand side of (13.10) is equal to

$$\int_B \mathbf{1}_A Y\,dP = \int \mathbf{1}_{B \cap A} Y\,dP = \int_{B \cap A} Y\,dP. \quad (\because B \cap A \in \mathcal{G}).$$

This proves that (13.10) holds for all $B \in \mathcal{G}$, and hence for all $\mathcal{G}$-measurable simple functions. Extend (13.10) to nonnegative functions and real-valued functions. ∎

> **Theorem 13.10**
>
> *Suppose X is a random variable in $(\Omega, \mathcal{F}, P)$ that is integrable and $\mathcal{G}$ is a sub-σ-algebra of $\mathcal{F}$. If $\sigma(X)$ and $\mathcal{G}$ are independent, then*
>
> $$E[X|\mathcal{G}] = E[X] \quad a.s.$$
>
> *In particular, when X and Y are independent random variables, we have $E[X|Y] = E[X]$ a.s.*

Proof It is sufficient to show that for each $B \in \mathcal{G}$,

$$\int_B X \, dP = \int_B E[X] \, dP. \tag{13.11}$$

Suppose $X = \mathbf{1}_A$ for some $\mathcal{F}$-measurable set A. For any set $B \in \mathcal{G}$, the left-hand side of (13.11) is

$$\int_B \mathbf{1}_A \, dP = \int \mathbf{1}_{A \cap B} \, dP = P(A \cap B) = P(A)P(B).$$

On the other hand, we can write the right-hand side of (13.11) as

$$\int_B E[\mathbf{1}_A] \, dP = E[\mathbf{1}_A] \int_B dP = P(A)P(B).$$

Therefore, (13.11) holds for $X = \mathbf{1}_A$ and any $B \in \mathcal{G}$.

We can extend this to simple functions, nonnegative functions, and finally to real-valued functions. ∎

Example 13.3.1 (Iterated Expectation)
For any random variables X and Y, apply Theorem 13.8 with $\mathcal{H} = \{\emptyset, \Omega\}$ and $\mathcal{G} = \sigma(Y)$. Since $E[Z|\mathcal{H}]$ is a constant and is equal to $E[Z]$ for any random variable Z, we get $E[E[X|Y]] = E[X]$.

13.4 Conditional Expectation Given a Discrete Random Variable

Although conditional expectation can be defined in a very general setting, we often want to compute it explicitly in concrete cases, such as when dealing with discrete random variable. To do so, we need the following property of Lebesgue integral.

Theorem 13.11
Suppose X is an integrable random variable defined on a probability space $(\Omega, \mathcal{F}, P)$ and $(A_i)_{i=1}^{\infty}$ be a sequence of mutually disjoint sets in $\mathcal{F}$. Then

$$\sum_{i=1}^{\infty} \int_{A_i} X \, dP = \int_{\cup_i A_i} X \, dP.$$

Proof For each integer $m \geq 1$, let $Y_m = X \sum_{i=1}^{m} \mathbf{1}_{A_i}$. We obviously have $|Y_m| \leq |X|$, and since $E[|X|]$ is finite by assumption, we can apply the dominated convergence theorem to obtain

$$\sum_{i=1}^{\infty} \int_{A_i} X \, dP = \lim_{m \to \infty} \sum_{i=1}^{m} \int_{\Omega} X \mathbf{1}_{A_i} \, dP = \int_{\Omega} \lim_{m \to \infty} \sum_{i=1}^{m} X \mathbf{1}_{A_i} \, dP = \int_{\cup_i A_i} X \, dP.$$

This proves the equality in Theorem 13.11. ∎

With the result in the previous theorem, we can compute the conditional expectation given a countable partition of the sample space.

Theorem 13.12
Suppose that X is an integrable random variable defined on a probability space $(\Omega, \mathcal{F}, P)$, and $A_i \in \mathcal{F}$, for $i = 1, 2, 3, \ldots,$ form a countable partition of Ω. Let $\mathcal{G}$ be the σ-algebra generated by $(A_i)_{i=1}^{\infty}$. Then, the conditional expectation $E[X|\mathcal{G}]$ is almost surely equal to $\sum_{i=1}^{\infty} E[X|A_i]\mathbf{1}_{A_i}$.

Proof Since the events A_i's are mutually disjoint and their union is Ω, the conditional expectation $E[X|\mathcal{G}]$ is a function that is constant in each A_i. Suppose $E[X|\mathcal{G}](\omega) = c_i$ for $\omega \in A_i$. We can determine the value of c_i from the defining property of conditional expectation

$$c_i P(A_i) = \int_{A_i} E[X|\mathcal{G}] \, dP = \int_{A_i} X \, dP.$$

Therefore, $c_i = \frac{1}{P(A_i)} \int_{A_i} X \, dP = E[X|A_i]$. We define a $\mathcal{G}$-measurable random variable

$$\hat{X}(\omega) \triangleq \sum_{i=1}^{\infty} E[X|A_i]\mathbf{1}_{A_i}(\omega).$$

By linearity of expectation, for any finite union $B = A_{i_1} \cup A_{i_2} \cup \cdots \cup A_{i_m}$, where $i_1, i_2, \ldots, i_m$ are m distinct indices, we have

$$\int_B \hat{X} \, dP = \sum_{k=1}^{m} \int_{A_{i_k}} \hat{X} \, dP = \sum_{k=1}^{m} \int_{A_{i_k}} X \, dP = \int_B X \, dP. \qquad (13.12)$$

For countably many indices, we can obtain the same equality by Theorem 13.11. This completes the proof that $\hat{X}$ is a version of $E[X|\mathcal{G}]$. $\blacksquare$

When the countable partition is induced by a discrete random variable, Theorem 13.12 can be rephrased as follows.

Theorem 13.13

Suppose X and Y are discrete random variables taking values in $\{0, 1, 2, \ldots, \}$. Let the joint pmf of X and Y be $p_{XY}(x, y)$, and the marginal distribution of Y be $p_Y(y) = \sum_{x \geq 0} p_{XY}(x, y)$. When $g(X)$ is integrable, we have

$$E[g(X)|Y = y] = \sum_{x \geq 0} g(x) p_{X|Y}(x|y),$$

where $p_{X|Y}(x|y)$ is defined as $\frac{p_{XY}(x,y)}{p_Y(y)}$. In particular, if X is integrable, the conditional expectation of X given Y is

$$E[X|Y = y] = \sum_{x \geq 0} x p_{X|Y}(x|y).$$

Proof Let A_y be the event $\{\omega : Y(\omega) = y\}$, for $y = 0, 1, 2, 3 \ldots$. From Theorem 13.12, for $\omega \in A_y$, we have

$$E[g(X)|Y](\omega) = \frac{1}{P(A_y)} \int_{A_y} g(x) \, dP = \frac{1}{p_Y(y)} \sum_{x \geq 0} g(x) p_{XY}(x, y).$$

Thus, we can write

$$E[g(X)|Y = y] = \sum_{x \geq 0} g(x) \frac{p_{XY}(x, y)}{p_Y(y)} = \sum_{x \geq 0} g(x) p_{X|Y}(x|y)$$

for $y = 0, 1, 2, \ldots$. $\blacksquare$

13.5 Conditional Expectation Given a Continuous Random Variable

While the process of deriving the formula for computing conditional expectation in the continuous case is similar to that in the discrete case, it is more complex due to the presence of events with probability zero.

Theorem 13.14

Suppose $(\mathbb{R}^2, \mathscr{B}(\mathbb{R}^2), P)$ is a probability space on $\mathbb{R}^2$, and X and Y denote the x- and y-coordinates, respectively. Let $g(x)$ denote a Borel function in $L^1(P)$.

Suppose the joint probability density function of X and Y is denoted by $f_{XY}(x, y)$, i.e., for any Borel set B in $\mathscr{B}(\mathbb{R}^2)$, we have

$$P(B) = \int_B f_{XY}(x, y)\, d\lambda(x, y),$$

where λ is the Lebesgue measure on $\mathbb{R}^2$. Then, we have

$$E[g(X)|Y] = \int_{\mathbb{R}} g(x) f_{X|Y}(x|y)\, d\lambda(x), \tag{13.13}$$

where

$$f_{X|Y}(x|y) \triangleq \frac{f_{XY}(x, y)}{f_Y(y)}$$

and

$$f_Y(y) \triangleq \int_{\mathbb{R}} f_{XY}(x, y)\, d\lambda(x). \tag{13.14}$$

In particular, if X is integrable, we can compute the conditional expectation $E[X|Y]$ by

$$E[X|Y] = \int_{\mathbb{R}} x f_{X|Y}(x, y)\, d\lambda(y).$$

We may assume $f_Y(y) \neq 0$ for simplicity so that we do not need to worry about division by zero. We note that the conditional expectation $E[g(X)|Y]$ in (13.13) is a $\sigma(Y)$-measurable function and hence is a function of y.

Proof Let $h(y) = \int_{\mathbb{R}} g(x) f_{X|Y}(x|y) \, d\lambda(x)$ be the candidate function for the conditional expectation. Our goal is to prove that, for any $\sigma(Y)$-measurable set B, the following equation holds:

$$\int_B g(x) \, dP(x, y) = \int_B h(y) \, dP(x, y). \tag{13.15}$$

Consider a special type of $\sigma(Y)$-measurable set of the form $B = \mathbb{R} \times [a, b]$ for some $a < b$. Assuming we can apply Fubini's theorem to evaluate the integrals by iterated integration, we will show that Eq. (13.15) holds for this type of set. When B is of the form $\mathbb{R} \times [a, b]$, we can express the right-hand side of (13.15) as

$$\int_{[a,b]} \int_{\mathbb{R}} h(y) f_{XY}(x, y) \, d\lambda(x) d\lambda(y) = \int_{[a,b]} h(y) \int_{\mathbb{R}} f_{XY}(x, y) \, d\lambda(x) d\lambda(y)$$

$$= \int_{[a,b]} h(y) f_Y(y) \, d\lambda(y)$$

$$= \int_{[a,b]} \int_{\mathbb{R}} g(x) f_{XY}(x, y) \, d\lambda(x) d\lambda(y).$$

The last line is the same as the left-hand side of (13.15).

We can justify the application of Fubini's theorem by assuming that $|g(x)|$ is integrable, as

$$\int_{\mathbb{R} \times [a,b]} |g(x)| f_{XY}(x, y) \, d\lambda(x, y) \leq \int_{\mathbb{R} \times \mathbb{R}} |g(x)| f_{XY}(x, y) \, d\lambda(x, y) < \infty.$$

We obtain the following equation for all closed intervals I in $\mathbb{R}$:

$$E[g(X)\mathbf{1}_{\mathbb{R} \times I}] = E[h(Y)\mathbf{1}_{\mathbb{R} \times I}]. \tag{13.16}$$

We can extend (13.16) to all Borel sets $B \in \mathscr{B}(\mathbb{R})$ by appealing to the π-λ theorem. First, we observe that the collection of all closed intervals in $\mathbb{R}$ form a π-system $\mathscr{P}$. Next, we consider the collection $\mathscr{L}$ consisting of subsets S of Ω that satisfies

$$E[g(X)\mathbf{1}_{\mathbb{R} \times S}] = E[h(Y)\mathbf{1}_{\mathbb{R} \times S}].$$

By what we have proved above, we know that $\mathscr{P} \subset \mathscr{L}$. We can prove that $\mathscr{L}$ is indeed a λ-system, which implies that $\mathscr{L}$ contains all Borel sets in $\mathbb{R}$. Therefore, (13.16) holds for all Borel sets $B \in \mathscr{B}(\mathbb{R})$.

Consequently, we have $E[g(X)\mathbf{1}_C] = E[h(Y)\mathbf{1}_C]$ for all $\sigma(Y)$-measurable sets C. This implies that $h(Y)$ is indeed a version of the conditional expectation $E[g(X)|Y]$, as it satisfies the defining properties of the conditional expectation. ∎

The second part of the theorem establishes the equivalence between the measure-theoretic definition of conditional expectation and the definition typically presented in a first course of probability for continuous random variables. By establishing this equivalence, we demonstrate that the abstract definition of conditional expectation generalizes the familiar notion from introductory probability theory.

While the classical approach to conditional expectation is well-suited for dealing with continuous or discrete random variables, it is not capable of handling the case where we wish to compute the conditional expectation of a discrete random variable given a continuous random variable, or vice versa. In such cases, the measure-theoretic definition provides a more general and rigorous framework for defining conditional expectation.

Example 13.5.1 (Conditional Probability Given a Continuous Random Variable)
Consider a point (X, Y) chosen uniformly at random in the interior of a unit square with vertices $(0, 0)$, $(0, 1)$, $(1, 1)$, and $(1, 0)$. Let Z be a binary random variable defined by

$$Z \triangleq \begin{cases} 1 & \text{if } Y < X \\ 0 & \text{if } Y \geq X. \end{cases}$$

We want to find $E[Z|X]$. Note that X is uniformly distributed between 0 and 1, while Z is a Bernoulli random variable.

Let A be the event $\{Y < X\}$. The random variable Z is the same as the indicator function $\mathbf{1}_A$, and $E[Z|X]$ is the conditional probability $P(A|X)$. From the geometry of the problem, we can make a guess that $P(A|X) = X$. Indeed, we will show that

$$\int_B Z \, dP = \int_B X \, dP \tag{13.17}$$

holds for all $\sigma(X)$-measurable sets B.

Consider $B = [a, b] \times [0, 1]$, which is $\sigma(X)$-measurable. We have

$$\int_{[a,b]\times[0,1]} Z \, dP = \int_{[a,b]\times[0,1]} \mathbf{1}_A \, dP = \frac{b^2 - a^2}{2}.$$

This is the area of a trapezoid with vertices $(a, 0)$, $(b, 0)$, (b, b), and (a, a).
On the other hand, we have

$$\int_B X \, dP = \int_a^b \int_0^1 x \, dy dx = \int_a^b x \, dx = \frac{b^2 - a^2}{2}.$$

Hence, (13.17) holds for all set B that can be written as a rectangle $[a, b] \times [0, 1]$ for some $a < b$. Using the π-λ theorem, we can extend this to all sets B that can be written as a Cartesian product $B = C \times [0, 1]$, where C is Borel set in $\mathbb{R}$. Finally, we note that X is $\sigma(X)$-measurable. Therefore, X is a version of the conditional expectation $E[\mathbf{1}_A|X]$, which is equal to $P(A|X)$ by the definition of conditional expectation. This proves that $E[Z|X] = P(A|X) = X$.

13.6   Application: Martingale and Stopping Time

The term "martingale" originated from French, where it referred to a half belt used to secure a horse or dog's neck. In probability theory, a martingale is a stochastic process that models a fair game in which the expected gain of a gambler is zero, assuming that the gambler can only make decisions in a causal manner. This probability model has found wide applications in finance.

We encode the information available from the history in the form of a filtration. (cf. Example 2.2.1)

Definition 13.6

A *filtration* is a sequence of increasing σ-subfields in $\mathscr{F}$, denoted as $(\mathscr{F}_n)_{n=0}^{\infty}$, such that

$$\mathscr{F}_0 \subseteq \mathscr{F}_1 \subseteq \mathscr{F}_2 \subseteq \cdots .$$

A sequence of random variables $(X_n)_{n=0}^{\infty}$ is said to be *adapted to* filtration $(\mathscr{F}_n)_{n=0}^{\infty}$ if X_n is $\mathscr{F}_n$-measurable for all n.

Another way to describe a filtration is to regard the index n as representing discrete time. At each time n, the filtration captures the information available up to that point in time. Typically, the first σ-field $\mathscr{F}_0$ in the filtration is the trivial σ-field $\{\emptyset, \Omega\}$, i.e., at time $n = 0$, only the trivial events $\emptyset$ and Ω are measurable.

The defining property of a filtration is that if something is measurable at time n, then it continues to be measurable at any future time $n' > n$. As time advances, we have more available information, and the σ-algebras in the filtration become larger.

Example 13.6.1 (Natural Filtration)
Suppose $Y_1, Y_2, \ldots$ is a sequence of random variables defined on probability space $(\Omega, \mathscr{F}, P)$. Let $\mathscr{F}_0 \triangleq \{\emptyset, \Omega\}$ and $\mathscr{F}_n \triangleq \sigma(Y_1, Y_2, \ldots, Y_n)$, for $n = 1, 2, 3, \ldots$. Then $(\mathscr{F}_n)_{n \geq 0}$ is a filtration. If $f_n(y_1, \ldots, y_n)$ is any $\mathscr{B}(\mathbb{R}^n)$-measurable function defined on $\mathbb{R}^n$, then the sequence $(f_n(Y_1, \ldots, Y_n))_{i=1}^{\infty}$ is adapted to $(\mathscr{F}_n)_{n \geq 0}$. This means that $f_n(Y_1, Y_2, \ldots, Y_n)$ depends only on the information available up to and including time n. This filtration is called the *natural filtration* for the sequence $(Y_i)_{i \geq 1}$.

Definition 13.7

A sequence of random variables $(X_n)_{n=0}^{\infty}$ is a *martingale* relative to filtration $(\mathscr{F}_n)_{n=0}^{\infty}$ if it satisfies the following three conditions:

1. $E[|X_n|] < \infty$ for all n, i.e., $X_n \in L^1(P)$.
2. $(X_n)_{n=0}^{\infty}$ is adapted to $(\mathscr{F}_n)_{n=0}^{\infty}$.
3. $E[X_{n+1}|\mathscr{F}_n] = X_n$ a.s. for $n \geq 0$.

The third condition captures the idea that the expected value of the next observation, given the current and all past observations, is equal to the current observation.

When the filtration is defined by a sequence of random variables $(Y_n)_{n \geq 0}$, then random variable X_n is a measurable function of $(Y_0, Y_1, \ldots, Y_n)$ (Theorem 12.6). We can express the last condition in Definition 13.7 as $E[X_{n+1}|Y_0, Y_1, \ldots, Y_n] = X_n$.

Martingale can be used to model a variety of stochastic processes, including a betting system, where X_n represents the amount of money in the gambler's pocket at time n. The essence of a martingale is that, regardless of the betting strategy adopted, the expected value of X_{n+1} in the next step, given the history up to time n, is always equal to the current value X_n. In other words, a martingale represents a fair game.

Theorem 13.15

In Definition 13.7, we can replace the third condition by

$$E[X_m|\mathscr{F}_n] = X_n \ a.s. \ for \ all \ m \geq n. \tag{13.18}$$

Proof To show that (13.18) implies the third condition in Definition 13.7, we can take $m = n + 1$. To prove the reverse direction, suppose $m \geq n + 1$. By the tower property (Theorem 13.8), we have $E[X_m|\mathscr{F}_n] = E[E[X_m|\mathscr{F}_{m-1}]|\mathscr{F}_n] = E[X_{m-1}|\mathscr{F}_n]$ a.s. Repeating this argument, we obtain $E[X_m|\mathscr{F}_n] = \cdots = E[X_{n+1}|\mathscr{F}_n] = X_n$ a.s. ∎

In a martingale, the unconditioned expectation $E[X_n]$ remains constant for all n.

Theorem 13.16

If $(X_n)_{n=0}^{\infty}$ is a martingale relative to some filtration $(\mathscr{F}_n)_{n=0}^{\infty}$, then $E[X_n] = E[X_0]$ for all $n \geq 1$.

Proof Take expectation on both sides of (13.18). ∎

Example 13.6.2 (Random Walk as Martingale)
Consider a sequence $(Y_n)_{n \geq 1}$ of i.i.d. random variables with zero mean, and define the partial sums $S_n \triangleq Y_1 + Y_2 + \cdots + Y_n$ for $n \geq 1$, with $S_0 \triangleq 0$. Let $\mathscr{F}_n \triangleq \sigma(Y_1, \ldots, Y_n)$ be the natural filtration for the sequence $(Y_n)_{n \geq 1}$ and set $\mathscr{F}_0 \triangleq \{\emptyset, \Omega\}$. It can be verified that $S_0, S_1, S_2, \ldots$ form a martingale relative to the filtration $(\mathscr{F}_n)_{n \geq 0}$.

To see this, we first note that S_n is a function of $Y_1, Y_2, \ldots, Y_n$ and is thus $\mathscr{F}_n$-measurable. Moreover, since Y_n has zero mean, we have

$$E[S_{n+1}|\mathscr{F}_n] = E[Y_1 + Y_2 + \cdots + Y_n|\mathscr{F}_n] + E[Y_{n+1}|\mathscr{F}_n]$$
$$= E[Y_1 + Y_2 + \cdots + Y_n|\mathscr{F}_n] + E[Y_{n+1}]$$
$$= Y_1 + Y_2 + \cdots + Y_n + 0 = S_n,$$

where the last equality follows from the independence of Y_{n+1} from $\mathscr{F}_n$ and the zero mean of Y_{n+1}. Finally, since $|S_n| \le \sum_{i=1}^{n} |Y_i|$, we have $E[|S_n|] \le \sum_{i=1}^{n} E[|Y_i|] < \infty$ for each n.

Example 13.6.3 (Doob's Martingale)

Suppose $Y_1, Y_2, \ldots$ is a sequence of random variables, and let X be an integrable random variable defined on the same probability space. Define the natural filtration $(\mathscr{F}_n)_{n \ge 0}$ by $\mathscr{F}_0 = \{\emptyset, \Omega\}$ and $\mathscr{F}_n = \sigma(Y_1, \ldots, Y_n)$, for $n \ge 1$. Let $Z_i \triangleq E[X|\mathscr{F}_i]$, for $i = 0, 1, 2, \ldots$.

We note that $Z_n = E[X|\mathscr{F}_n]$ is $\mathscr{F}_n$-measurable by the definition of conditional expectation. By the triangle inequality of conditioned expectation, we have

$$E[|Z_n|] = E[|E[X|\mathscr{F}_n]|] \le E[E[|X| \,|\mathscr{F}_n]] = E[|X|] < \infty.$$

Finally, for $n \ge 1$, we can apply the tower property to obtain

$$E[Z_n|\mathscr{F}_{n-1}] = E[E[X|\mathscr{F}_n]|\mathscr{F}_{n-1}] = E[X|\mathscr{F}_{n-1}] = Z_{n-1}.$$

It proves that the sequence $(Z_n)_{n=0}^{\infty}$ is a martingale sequence relative to $(\mathscr{F}_n)_{n \ge 0}$.

The stopping time T is a device that terminates a martingale. It is a random variable such that the event $\{T(\omega) = n\}$ can be determined by the information up to time n. For example, a constant T is always a stopping time, but the first time that a gambling loses ten times in a row is not, since it depends on future outcomes that are not known at time n.

Definition 13.8

Given a filtration $(\mathscr{F}_n)_{n=0}^{\infty}$ and a random variable

$$T : \Omega \to \{0, 1, 2, \ldots, \} \cup \{\infty\}$$

defined on $(\Omega, \mathscr{F}, P)$, we say that T is a *stopping time* relative to the filtration $(\mathscr{F}_n)_{n=0}^{\infty}$ if the event $\{T \le n\}$ is $\mathscr{F}_n$-measurable for all n.

If $(X_n)_{n=0}^{\infty}$ is a sequence of random variables and T is a stopping time relative to $(\sigma(X_0, \ldots, X_n))_{n \ge 0}$, we define a random variable X_T by

$$X_T(\omega) \triangleq X_{T(\omega)}(\omega).$$

The random variable X_T represents the value of the sequence $(X_n)_{n \ge 0}$ at the stopping time T, which is the time at which the martingale is terminated. Formally, for each $\omega \in \Omega$, we define $X_T(\omega) \triangleq X_{T(\omega)}(\omega)$, where $T(\omega)$ is the time at which the stopping rule is applied to the realization ω.

Example 13.6.4 (Example of Stopping Time)
Let $(X_n)_{n=0}^{\infty}$ be a sequence of random variables and let $(\mathscr{F}_n)_{n=0}^{\infty}$ be the filtration defined by $\mathscr{F}_n = \sigma(X_0, X_1, \ldots, X_n)$. Define a random variable T as the first time n that the values of X_n exceed 100. Mathematically, we write

$$T(\omega) = \begin{cases} \min\{n : X_n(\omega) \geq 100\} & \text{if } X_n(\omega) \geq 100 \text{ for some } n \\ \infty & \text{otherwise.} \end{cases}$$

When the sequence $(X_n(\omega))_{n\geq 1}$ never exceeds the threshold 100, we define $T(\omega)$ to be infinity for this realization. The random variable T is a stopping time relative to filtration $(\mathscr{F}_n)_{n=0}^{\infty}$ because the event $\{\omega : T(\omega) \leq n\}$ can be written as the union of $\{\omega : X_i(\omega) \geq 100\}$, for $i = 0, 1, \ldots, n$, which are all $\mathscr{F}_n$-measurable.

Definition 13.9

Given a martingale $(X_n)_{n\geq 0}$ and a stopping time T, a *stopped process* $(X_n^T)_{n\geq 0}$ is obtained from $(X_n)_{n\geq 0}$ by forcing the values with indices larger than T to be X_T. That is, $X_n^T(\omega) \triangleq X_{T \wedge n}(\omega)$.

For all $m \geq T$, the random variable X_m^T stops updating and is equal to X_T. In the definition, we used the notation $a \wedge b \triangleq \min(a, b)$.

Theorem 13.17
The stopped process $(X_n^T)_{n\geq 0}$ is a martingale relative to the filtration $(\mathscr{F}_n)_{n\geq 0}$, and $E[X_n^T] = E[X_0]$ for all $n \geq 0$.

Proof The expectation $E[X_n^T]$ is less than or equal to $\max_k E[X_k]$, with maximum taken over $k = 0, 1, \ldots, n$. Since $E[X_k]$ is finite for all k, the maximum is also finite.

For $n \geq 0$, the n-th random variable X_n^T in the stopped process is a function of $X_0, X_1, \ldots, X_n$ and T. Since T is a stopping time, X_n^T is $\mathscr{F}_n$-measurable.

We write X_{n+1}^T as $X_n^T + (X_{n+1} - X_n)\mathbf{1}_{\{T>n\}}$. The conditional expectation of X_{n+1}^T given $\mathscr{F}_n$ is

$$E[X_{n+1}^T | \mathscr{F}_n] = X_n^T + \mathbf{1}_{\{T>n\}} E[X_{n+1} - X_n | \mathscr{F}_n] = X_n^T + \mathbf{1}_{\{T>n\}} \cdot 0 = X_n^T.$$

This proves that X_n^T is a martingale relative to $(\mathscr{F}_n)_{n\geq 0}$.

Because $(X_n^T)_{n\geq 1}$ is a martingale, the last statement about the expectation $E[X_n^T]$ follows from Theorem 13.16. ∎

In contrast to the previous theorem, we are not just interested in the expected value of the stopped process at a particular time, but also interested in the expected value of the martingale sequence when it stops. The following theorem is called the martingale stopping theorem, which is also known as Doob's optional stopping theorem. Under the gambling scenario, the first condition in the theorem means that a gambler has to leave the casino within a fixed duration. The second one requires that the game must end when the gambler lose all of his fortune or win a predetermined amount of money. The third one models the scenario in which the increments are uniformly bounded.

Theorem 13.18 (Martingale Stopping Theorem)

Let $(X_n)_{n=0}^{\infty}$ be a martingale and T be a stopping time relative to filtration $(\mathscr{F}_n)_{n=0}^{\infty}$. Then

$$E[X_T] = E[X_0]$$

if one of the following thee conditions hold:

(i) T is bounded by a constant C almost surely.
(ii) T is almost surely finite, and there exists a constant K such that $|X_n| \leq K$ for all n.
(iii) $E[T] < \infty$, and there exists a constant c such that $|X_n - X_{n-1}| \leq c$ for all n.

Proof Under conditions (i), (ii), or (iii), the stopping time T is finite with probability 1. Therefore, X_T is well-defined almost surely and $X_{T \wedge n} \to X_T$ a.s. as $n \to \infty$.

(i) The martingale is stopped before time C a.s. We have $X_T = X_C$ a.s. Hence, $E[X_T] = E[X_C] = E[X_0]$.
(ii) Since $|X_{T \wedge n}|$ is bounded by a constant for all n, by applying the dominated convergence theorem, we obtain

$$\lim_{n \to \infty} E[X_{T \wedge n}] = E[X_{\lim_n (T \wedge n)}] = E[X_T].$$

In the second equality, we have used the assumption that T is finite a.s. and thus $\lim_n (T \wedge n) = T$ a.s. From Theorem 13.17, we know that $E[X_{T \wedge n}] = E[X_n^T] = E[X_0]$ for all n. Therefore, $E[X_T] = E[X_0]$.
(iii) Suppose $E[T]$ is finite, and there exists a constant c such that $|X_n - X_{n-1}| \leq c$ for all n. We write $X_{T \wedge n}$ as a telescoping sum

$$X_{T \wedge n} = X_0 + \sum_{k=1}^{T \wedge n} (X_k - X_{k-1}).$$

The number of terms is random but is no larger than n with probability 1. Hence, we are dealing with finitely many additions in the calculation of $X_{T \wedge n}$. Using the bounded difference assumption, we obtain

$$|X_{T \wedge n}| \leq |X_0| + \sum_{k=1}^{T \wedge n} |X_k - X_{k-1}| \leq |X_0| + c \cdot (T \wedge n) \leq |X_0| + cT.$$

The right-hand side is integrable because both X_0 and T are integrable. By the dominated convergence theorem, we conclude that $\lim_{n \to \infty} E[X_{T \wedge n}] = E[X_T]$. Since $E[X_{T \wedge n}] = E[X_0]$ for all n by Theorem 13.17, we obtain $E[X_T] = E[X_0]$.   ∎

The next example is commonly known as the ABRACADABRA problem [10].

Example 13.6.5 (Martingale Patterns and Gambling Team)

Suppose a monkey type randomly on a keyboard with only three keys: a, b, and c. Two players bet on whether the pattern "abab" or "aabb" will appear first in the resulting random string. We analyze who will win this game by computing the expected waiting time until we see each pattern. We solve the prolem by introducing a casino. At each round, a random letter is drawn. A person can bet m on the random outcome. He/she will receive $3m$ if the guess is correct, but will lose all of the bet otherwise. This makes it a fair game.

We first consider the pattern "abab". We introduce a team of gamblers, who enter the game one at a time. Gambler i bets \$1 at time i that the i-th letter is the character "a". This gambler will receive \$3 if the letter "a" is drawn at time i but will leave otherwise. Then he will bet \$3 to bet the next drawn letter is "b". He will receive \$9 if the letter at time $i + 1$ is indeed "b" but will leave otherwise. The gambler will then bet \$27 at time $i + 2$ for the letter "a" and \$81 at time $i + 3$ for the letter "b". He will leave the game whenever he loses.

Assume that there are infinitely many potential gamblers waiting outside, and at each time, a new gambler enters the game. We stop the game when a gambler has just won four times in a row. This forms a stopped martingale satisfying the third condition in Theorem 13.18. Suppose the game is stopped at time T. At this time, exactly two gamblers are still in the game. The gambler who entered at time $T - 1$ wins \$9 −\$1 = \$8, and the one who entered at time $T - 3$ wins \$81−\$1=\$80. All other gamblers lose \$1. By Theorem 13.18, the expected gain of the whole gambling team is 0,

$$0 = E[X_0] = E[X_T] = E[(T - 2)(-\$1) + (\$81 - \$1) + (\$9 - \$1)] = \$90 - \$E[T].$$

Therefore, the expected waiting time for the string "abab" is 90.

If we change the pattern to "aabb", only one gambler is in the game when we first see the pattern. A similar calculation shows that the expected waiting time is 81.

$$0 = E[X_0] = E[X_T] = E[(T - 1)(-\$1) + \$81 - \$1] = \$81 - \$E[T].$$

We simulate this experiment by the following Python program.

```python
from random import choice

A = ['a','b','c']                      # alphabet = {a,b,c}

def monkey(alphabet,pattern):
    random_string = ""                 # initialize to empty string
    while True:
        random_string += choice(alphabet) # randomly draw a character
        if random_string[-len(pattern):] == pattern:
            break                      # break if we see the pattern at the end
    return len(random_string)          # return the length of the random string

n = 10000   # run the experiment n times
```

```
waiting_time1 = sum([monkey(A,'abab') for _ in range(n)])/n
waiting_time2 = sum([monkey(A,'aabb') for _ in range(n)])/n

print(f"Average waiting time for pattern abab = {waiting_time1}")
print(f"Average waiting time for pattern aabb = {waiting_time2}")
```

The function monkey generates a random string that ends with the given pattern and returns the length of the random string. We run 10,000 experiments for each of the two patterns "abab" and "aabb" and compute the average length of the random string. Note that the only difference between the two experiments is the ending patterns. A sample output of the program is

```
Average waiting time for pattern abab = 89.1899
Average waiting time for pattern aabb = 81.2196
```

The simulation results are in accordance with the theoretic analysis.

Problems

13.1. Random variables X and Y take values in $\{1, 2, 3\}$. Their joint pmf is defined as in the following table.

	$Y = 1$	$Y = 2$	$Y = 3$
$X = 1$	0.1	0	0.2
$X = 2$	0	0.3	0
$X = 3$	0.2	0.2	0

Find $E[X|Y]$ and $E[Y|X]$.

13.2. Let X be a zero-mean Gaussian random variable with nonzero variance σ^2. Let

$$Y \triangleq \begin{cases} -1 & \text{if } X < -\sigma \\ 0 & \text{if } -\sigma \le X \le \sigma \\ 1 & \text{if } X > \sigma. \end{cases}$$

Compute the conditional expectation $E[X|Y]$ and $E[X|Y^2]$.

13.3. The joint pdf of two jointly Gaussian random variables X_1 and X_2 is given in (1.3). Find the conditional expectation of X_2 given $X_1 = x$.

13.4. Suppose X and Y are jointly Gaussian random variables. Find the conditional expectation of X given Y, in terms of the means, variances, and the covariance of X and Y.

13.5. (a) Suppose X and Y are i.i.d. random variables with finite mean. Prove that $E[X|X+Y] = (X+Y)/2$. (Hint: Show that $E[Y|X+Y] = E[X|X+Y]$.)
(b) Extend the result in part (a) to n i.i.d. random variables.

13.6. Prove the conditional version of Jensen inequality: $\phi(E[X|\mathcal{G}]) \leq E[\phi(X)|\mathcal{G}]$ for convex function $\phi(x)$.

13.7. Suppose X and Y are random variables defined on the same probability space and $E[X^2] < \infty$.

(a) Prove the identity $\text{Var}(X) = \text{Var}(E[X|Y]) + E[(X - E[X|Y])^2]$.
(b) Show that the last term in part (a) can be interpreted as

$$E[(X - E[X|Y])^2] = E[E[X^2|Y] - E[X|Y]^2].$$

13.8. (Generalized Bayes rule) Let $(\Omega, \mathcal{G}, P)$ be a probability space and $\mathcal{G}$ be a sub-σ-algebra of $\mathcal{F}$. For any $B \in \mathcal{G}$ and $A \in \mathcal{F}$ with $P(A) \neq 0$, deduce that

$$\frac{\int_B P(A|\mathcal{G})\,dP}{\int_\Omega P(A|\mathcal{G})\,dP} = \frac{P(B \cap A)}{P(A)}.$$

Show that this equation reduces to the classical Bayes rule when $\mathcal{G}$ is generated by a partition of Ω.

13.9. Suppose $(X_n)_{n \geq 0}$ is a sequence of random variables defined on a probability space (we take $X_0 = 0$ as the initial r.v.). For $n \geq 1$, let

$$Y_n \triangleq X_n - E[X_n|X_0, X_1, \ldots, X_{n-1}].$$

The random variable Y_n can be interpreted as the new information contained in X_n relative to the information contained in the past $X_0, \ldots, X_{n-1}$. Let S_n denote the partial sum $S_n \triangleq Y_1 + Y_2 + \cdots + Y_n$ for $n \geq 1$ and $S_0 = 0$. Show that $(S_n)_{n \geq 0}$ is a martingale relative to the filtration $(\sigma(X_1, X_2, \ldots, X_n))_{n \geq 0}$.

13.10. Let X_n's be i.i.d. random variables with finite mean $\mu = E[X_1]$:

(a) Let T be an integer-valued random variable independent of $(X_n)_{n \geq 1}$, and assume T has finite expectation. Prove Wald's identity

$$E[X_1 + X_2 + \cdots + X_T] = \mu \cdot E[T].$$

(b) Prove that Condition (iii) in Theorem 13.18 can be relaxed to $E[T] < \infty$ and $E[|X_n - X_{n-1}| \, |\mathcal{F}_n] \leq c$ for all n.

(c) Let T be a stopping time for the filtration $\mathscr{F}_n \triangleq \sigma(X_1, \ldots, X_n)$, for $n \geq 1$. Apply part (b) to prove that Wald's identity holds if $E[|X_1|] < \infty$ and $E[T] < \infty$.

13.11 (Polya's Urn Model). Consider an urn containing b black balls and w white balls at time 0. At time i (for $i \in \mathbb{N}$), we pick a ball uniformly at random from the urn. If the color of the ball is black, we put two black balls back to the urn. If the color is white, we put two white balls back. There are $b + w + i$ balls in the urn after step i. Let B_i denote the number of black balls in the urn after step i:

(a) Compute the probabilities $P(B_1 = b)$ and $P(B_1 = b + 1)$.
(b) Let $\mathscr{F}_i$ be the σ-algebra generated by $B_0, B_1, \ldots, B_i$, for $i \geq 1$. Let $Z_i = B_i/(b + w + i)$ be the fraction of black balls in the urn after the i-th step. Show that $(Z_i)_{i \geq 0}$ is a martingale relative to the filtration $(\mathscr{F}_i)_{i \geq 0}$. (We define $B_0 = b$, $Z_0 = b/(b + w)$, and $\mathscr{F}_0 = \{\emptyset, \Omega\}$.)
(c) Compute $E[Z_i]$ for $i \geq 1$.

Levy's Continuity Theorem and Central Limit Theorem

14

The characteristic function of a random variable is a powerful tool that encapsulates all the information about its probability distribution. Lévy's continuity theorem establishes the equivalence between pointwise convergence of characteristic functions and convergence in distribution. This theorem offers a convenient way to determine whether a sequence of random variables converges in distribution, and serves as a tool for proving the central limit theorem.

Theorem 14.1 (Lévy's Continuity Theorem)
Let $(X_n)_{n=1}^{\infty}$ be a sequence of real-valued random variables. Let P_n denote the probability measure induced by X_n on $\mathbb{R}$, and $\phi_n(t)$ be the characteristic function of X_n, for $n \geq 1$. Suppose the characteristic functions converge pointwise to some function $\phi(t)$.

Then the followings are equivalent:

1. *X_n converges weakly to some random variable X as n tends to infinity.*
2. *$\phi(t)$ is the characteristic function of a random variable.*
3. *$\phi(t)$ is continuous at $t = 0$.*
4. *The sequence of probability measures $(P_n)_{n=1}^{\infty}$ is tight.*

In the first section in this chapter, we define the notion of weak convergence, which is equivalent to convergence in distribution. We also prove that convergence in total variation implies convergence in distribution. Then, we introduce the notion of tightness of measures and Prokhorov theorem. In the last section, we use the continuity theorem to prove a version of central limit theorem that assumes i.i.d. random variables with finite variance. In the last section, we will state a central limit theorem for triangular array and illustrate it through an example.

© The Author(s), under exclusive license to Springer Nature Switzerland AG 2023   237
K. Shum, *Measure-Theoretic Probability*, Compact Textbooks in Mathematics,
https://doi.org/10.1007/978-3-031-49830-5_14

14.1   Weak Convergence

Suppose we can only obtain information about a probability distribution P through some measurements of the form $\int h \, dP$, where $h(x)$ is a "test function". If we have a sequence of probability distributions P_n's, it would be desirable if, for each test function $h(x)$, the sequence of measurements $\int h \, dP_n$ converges as n tends to infinity. This motivates the concept of weak convergence, which allows us to study the convergence of probability distributions based on their behavior with respect to test functions.

Definition 14.1

Let P_n, for $n \geq 1$, be probability measures and P be another probability measure defined on sample space $\mathbb{R}$. If

$$\lim_{n \to \infty} \int_{\mathbb{R}} h(x) \, dP_n(x) = \int_{\mathbb{R}} h(x) \, dP(x)$$

for all bounded and continuous function $h(x)$, then we say that P_n *converges weakly* to P. The notation for weak convergence is

$$P_n \xrightarrow{W} P \text{ or } P_n \Rightarrow P.$$

We often apply this definition with the probability measures are the push-forward measures of some random variables. By using the change-of-variable formula (Theorem 7.11), we can formulate weak convergence for a sequence of random variables as follows.

Definition 14.2

Suppose $X_1, X_2, X_3 \ldots$, is a sequence of random variables, and X is a random variable. We say that $(X_n)_{n=1}^{\infty}$ *converges weakly* to X if the image measures of X_n's converge weakly to the image measure of X, i.e.,

$$\lim_{n \to \infty} E[h(X_n)] = E[h(X)]$$

for all continuous and bounded functions h.

The next theorem establishes the equivalence of weak convergence and convergence in distribution. Weak convergence is an elegant definition, and it is more convenient for proofs. However, checking the condition for weak convergence can be challenging, as it involves considering infinitely many potential test functions. In contrast, the definition of convergence in distribution is often more practical to use in calculations.

Theorem 14.2 (Weak Convergence and Convergence in Distribution)
Suppose $(X_n)_{n=1}^{\infty}$ is a sequence of random variables, and X be another random variable. Denote the cumulative distribution function of X_n by $F_n(x)$, for $n \geq 1$, and the cumulative distribution function of X by $F(x)$. We have

$$X_n \xrightarrow{W} X \text{ if and only if } F_n \xrightarrow{D} F.$$

Proof ($\Rightarrow$) Suppose $F(x)$ is continuous at $x = a$, i.e., $P(\{a\}) = 0$. The idea of proof is to approximate an indicator function by piece-wise linear function. Define a piece-wise linear function g by

$$g(x) = \begin{cases} 1 & \text{if } x < a - \epsilon, \\ \frac{1}{\epsilon}(a - x) & \text{if } a - \epsilon \leq x \leq a, \\ 0 & \text{if } x > a. \end{cases}$$

The function g is linearly decreasing from 1 to 0 in the interval $[a - \epsilon, a]$ and is continuous and bounded. We have

$$\mathbf{1}_{(-\infty, a-\epsilon]}(x) \leq g(x) \leq \mathbf{1}_{(-\infty, a]}(x).$$

Substitute x by X and X_n in the above inequalities, and take expectation to get

$$F(a - \epsilon) \leq E[g(X)] = \lim_{n \to \infty} E[g(X_n)] \leq \liminf_{n \to \infty} F_n(a). \tag{14.1}$$

The last inequality in (14.1) follows from $E[g(X_n)] \leq F_n(a)$ for all n.

Similarly, consider the function $h(x) = g(x - \epsilon)$. The function $h(x)$ satisfies

$$\mathbf{1}_{(-\infty, a]}(x) \leq h(x) \leq \mathbf{1}_{(-\infty, a+\epsilon]}(x)$$

and is continuous and bounded. We thus obtain

$$\limsup_{n \to \infty} F_n(a) \leq \lim_{n \to \infty} E[h(X_n)] = E[h(X)] \leq F(a + \epsilon). \tag{14.2}$$

Putting (14.1) and (14.2) together, we get

$$F(a - \epsilon) \leq \liminf_{n \to \infty} F_n(a) \leq \limsup_{n \to \infty} F_n(a) \leq F(a + \epsilon).$$

Since $F(x)$ is continuous at $x = a$, the limit of $(F_n(a))_{n \geq 1}$ exists and equals $F(a)$.

($\Leftarrow$) We apply Skorokhod representation theorem (Theorem 10.8), and let $(Y_n)_{n\geq 1}$ be a sequence of random variables defined on the same probability space, such that Y_n and X_n have the same distribution and Y_n's converge almost surely to a random variable Y that has the same distribution as X.

Let $h(y)$ be a continuous and bounded function on $\mathbb{R}$. Because h is continuous, by the continuous mapping theorem (Theorem 10.11), $h(Y_n)$ converges to $h(Y)$ almost surely. Since $h(Y_n)$ is bounded, we can apply the dominated convergence theorem to obtain

$$\lim_{n\to\infty} E[h(X_n)] = \lim_{n\to\infty} E[h(Y_n)] = E[\lim_{n\to\infty} h(Y_n)] = E[h(Y)] = E[h(X)].$$

Because it holds for any continuous and bounded function $h(x)$, this proves that X_n converges weakly to X. ∎

We can now prove that Condition 1 implies Condition 2 in Theorem 14.1.

Theorem 14.3
If $X_n \xrightarrow{W} X$, then $\phi_{X_n}(t)$ converges pointwise to $\phi_X(t)$ for all t.

Proof The real and imaginary parts of e^{itx} are $\cos(tx)$ and $\sin(tx)$, which are continuous and bounded. Therefore $E[\cos(tX_n)] \to E[\cos(tX)]$ and $E[\sin(tX_n)] \to E[\sin(tX)]$ as n approaches infinity. Therefore $E[e^{itX_n}]$ converges to $E[e^{itX}]$ as $n \to \infty$. ∎

Using the equivalent conditions in Theorem 14.2, we can show that convergence in total variation implies weak convergence and thus complete the proof of Theorem 10.6.

Theorem 14.4
If probability measures μ_n, for $n = 1, 2, 3, \ldots$, defined on the measurable space $(\mathbb{R}, \mathscr{B}(\mathbb{R}))$ converge in total variation distance, then they converge weakly.

Proof Suppose that $(\mu_n)_{n=1}^{\infty}$ converges in total variation to μ, i.e., $d_{TV}(\mu_n, \mu) \to 0$ as $n \to \infty$. For any fixed index n, define a measure $\nu_n \triangleq (\mu_n + \mu)/2$. We have $\mu_n \ll \nu_n$ and $\mu \ll \nu_n$. By applying Radon–Nikodym theorem (Theorem 13.5), we can represent μ_n by density f_n, and μ by density f, relative to the measure ν_n.

Let $h : \mathbb{R} \to \mathbb{R}$ denote a bounded and continuous function, and $h_{\max}$ be an upper bound of $h(x)$ over all x. By triangle inequality (Theorem 7.1), we get

$$\left| \int_{\mathbb{R}} h(x) \, d\mu_n(x) - \int_{\mathbb{R}} h(x) \, d\mu(x) \right| = \left| \int_{\mathbb{R}} h(x) f_n(x) \, dv_n(x) \right.$$

$$\left. - \int_{\mathbb{R}} h(x) f(x) \, dv_n(x) \right|$$

$$\leq h_{\max} \int_{\mathbb{R}} |f_n(x) - f(x)| \, dv_n(x).$$

The proof of Theorem 8.6 can be adapted to probability measures that are absolutely continuous with respect to any probability measure. Indeed, we can use similar argument to prove that

$$d_{TV}(\mu_n, \mu) = \frac{1}{2} \int_{\mathbb{R}} |f_n(x) - f(x)| \, dv_n(x),$$

which yields

$$\left| \int_{\mathbb{R}} h(x) \, d\mu_n(x) - \int_{\mathbb{R}} h(x) \, d\mu(x) \right| \leq 2 h_{max} d_{TV}(\mu_n, \mu).$$

Since $d_{TV}(\mu_n, \mu)$ converges to 0, we also have $\int h \, d\mu_n \to \int h \, d\mu$ as $n \to \infty$.

The function $h(x)$ can be any continuous and bounded function in the previous paragraphs. This proves that probability measure μ_n converges weakly to μ.   ∎

We note that the converse of Theorem 14.4 does not hold in general. This is because the total variation distance between a discrete distribution and a continuous distribution is always equal to 1. For instance, the total variation distance between a binomial distribution and a normal distribution is equal to 1. However, there exist examples where binomial distribution converges in distribution to the normal distribution, despite having a total variation distance of 1 between them. (see example 10.3.1)

14.2   Tightness of a Sequence of Measures

The condition of tightness prevents the probability escaping to infinity in a sequence of measures.

Definition 14.3

A sequence of probability measures $(P_n)_{n=1}^{\infty}$ on $\mathbb{R}$ is said to be *tight* if for all $\epsilon > 0$, there exists a sufficiently large M_ϵ such that

$$P_n([-M_\epsilon, M_\epsilon]) > 1 - \epsilon \qquad \text{for all } n.$$

Similarly, a sequence of cdf $(F_n)_{n=1}^{\infty}$ is said to be *tight* if for any $\epsilon > 0$ there exists M_{ϵ} such that $F_n(M_{\epsilon}) - F_n(-M_{\epsilon}) > 1 - \epsilon$ for all n.

Example 14.2.1 (An Example of Non-tight Sequence of Distributions)
For $n \geq 1$, let P_n be the uniform distribution on the interval $[-n, n]$. The pdf of P_n is $f_n(x) = (2n)^{-1}\mathbf{1}_{[-n,n]}(x)$. For any positive real number M, the probability $P_n([-M, M])$ converges to 0 as $n \to \infty$. This sequence of probability measures is not tight.

The following theorem is often applied when P_n is the probability distribution induced by a random variable X_n. It provides a sufficient condition for checking tightness using characteristic functions.

Theorem 14.5
Let $(P_n)_{n=1}^{\infty}$ be a sequence of distribution on $\mathbb{R}$, and let $\phi_n(t)$ denote the characteristic function of distribution P_n. If $(\phi_n(t))_{n=1}^{\infty}$ converges to a function $c(t)$ that is continuous at $t = 0$, then $(P_n)_{n=1}^{\infty}$ is tight.

Proof Because $\phi_n(0) = 1$ for all n, we have $c(0) = 1$. Moreover, note that $1 - \phi_n(t) = 0$ when $t = 0$. The idea of proof is to examine the behavior of $1 - \phi_n(t)$ when t is close to zero. By assumption, we know that $1 - \phi_n(t)$ converges to the function $1 - c(t)$, which is continuous at $t = 0$.

The first step is to apply Fubini theorem to derive

$$\int_{-u}^{u} (1 - \phi_n(t)) \, dt = \int_{\mathbb{R}} \int_{-u}^{u} (1 - e^{itx}) \, dt \, dP_n(x) \tag{14.3}$$

for any positive u. The order of integration can be exchanged because the integrand has absolute value bounded by 2.

Next, for any $u > 0$ and $x \neq 0$, the inner integral can be computed using

$$\int_{-u}^{u} 1 - e^{itx} \, dt = 2 \int_{0}^{u} 1 - \cos tx \, dt = 2\left(u - \frac{\sin ux}{x}\right). \tag{14.4}$$

Combining (14.3) and (14.4), we obtain

$$\frac{1}{u} \int_{-u}^{u} (1 - \phi_n(t)) \, dt = \int_{\mathbb{R}} 2\left(1 - \frac{\sin ux}{ux}\right) dP_n(x). \tag{14.5}$$

On the other hand, for $|x| \geq 2/u$, we can bound the integral in the second integral above as follows:

$$2\left(1 - \frac{\sin(ux)}{ux}\right) = 2\left(1 - \frac{\sin(u|x|)}{u|x|}\right) \geq 2\left(1 - \frac{1}{u|x|}\right) \geq 1.$$

Therefore,

$$2\left(1 - \frac{\sin ux}{ux}\right) \geq \begin{cases} 1 & \text{for } |x| \geq \frac{2}{u} \\ 0 & \text{for } |x| < \frac{2}{u}. \end{cases}$$

Here, we use the fact that $\sin(u|x|) < u|x|$ when $u|x| < 2$. Consequently, for any $u > 0$,

$$\left| \frac{1}{u} \int_{-u}^{u} (1 - \phi_n(t))\, dt \right| \geq P_n\left(|x| \geq \frac{2}{u}\right). \tag{14.6}$$

The remaining task is to find a sufficiently small u, such that the left-hand side of (14.6) is bounded by a constant for all n. Applying the triangle inequality for real numbers, we can write

$$\left| \frac{1}{u} \int_{-u}^{u} (1 - \phi_n(t))\, dt \right| = \left| \frac{1}{u} \int_{-u}^{u} (1 - c(t) + c(t) - \phi_n(t))\, dt \right|$$

$$\leq \frac{1}{u} \int_{-u}^{u} |1 - c(t)|\, dt + \frac{1}{u} \int_{-u}^{u} |c(t) - \phi_n(t)|\, dt. \tag{14.7}$$

The first integral in (14.7) can be made arbitrarily small because it is assumed that $c(t)$ is continuous at $t = 0$. We can pick a sufficiently small $u_0 > 0$ such that

$$\frac{1}{u_0} \int_{-u_0}^{u_0} |1 - c(t)|\, dt \leq \epsilon/2.$$

For the second term in (14.7), we use the assumption that $\phi_n(t)$ is converging to $c(t)$ to show that

$$\lim_{n \to \infty} \frac{1}{u_0} \int_{-u_0}^{u_0} |c(t) - \phi_n(t)|\, dt = \frac{1}{u_0} \int_{-u_0}^{u_0} \lim_{n \to \infty} |c(t) - \phi_n(t)|\, dt = 0.$$

Because the integrand is bounded by a constant 2, we can apply the dominated convergence theorem and choose a sufficiently large integer N such that

$$\frac{1}{u_0} \int_{-u_0}^{u_0} |c(t) - \phi_n(t)|\, dt \leq \epsilon/2$$

for all $n \geq N$. Consequently, we obtain

$$P_n\left(|x| \geq \frac{2}{u_0}\right) \leq \epsilon$$

for $n \geq N$.

Since there are only finitely many positive integers that are less than N, for each $i = 1, 2, \ldots, N-1$, we can pick M_i such that $P_i(|x| \geq M_i) \leq \epsilon$. Finally, we choose a value of M that is larger than $M_1, M_2, \ldots, M_{N-1}$, and $2/u_0$. Then $P_n(|x| > M) \leq \epsilon$ for all n. This completes the proof that the sequence $(P_n)_{n=1}^{\infty}$ is tight. ∎

14.3   Prokhorov Theorem and Sequential Compactness

In the previous two sections, we proved that Condition 1 implies Condition 2, and Condition 3 implies Condition 4 in Lévy's continuity theorem. The reason why Condition 2 implies Condition 3 is characteristic function's continuity (Theorem 9.3). It remains to prove that Condition 1 follows from Condition 4. The final step requires the Prokhorov theorem, which we state without proof as follows.

> **Theorem 14.6 (Prokhorov)**
>
> *If $(F_n)_{n=1}^{\infty}$ is a sequence of tight cumulative distribution functions, then there exists a subsequence converging in distribution to the cumulative distribution of a probability measure.*

The Prokhorov theorem is analogous to the Bolzano–Weierstrass theorem. In real analysis, a subset S in $\mathbb{R}^n$ is said to be sequentially compact if any infinite sequence in S has a convergent subsequence that converges to an element in S. The Prokhorov theorem says that the space of tight distributions with topology defined by weak convergence is sequentially compact. For a proof of this theorem, see [4, Thm 3.2.12 and 3.2.13].

We now complete the proof of Levy's theorem. Consider a sequence $(P_n)_{n\geq 1}$ of probability measure that is a tight, and let $(F_n)_{n\geq 1}$ be the corresponding cumulative distribution functions. In Lévy's continuity theorem, the characteristic function $\phi_n(t)$ of P_n is assumed to converge pointwise, with the limit function denoted by $\phi(n)$.

By the Prokhorov theorem, there exists a subsequence $(a_n)_{n\geq 1}$ of $\mathbb{N}$ such that $F_{a_n}(x)$ converges pointwise to the cumulative distribution function $F(x)$ of a probability measure P at every continuity point of $F(x)$. Using Theorem 14.3 and the equivalence of weak convergence and convergence in distribution, we can show that the characteristic functions $\phi_{a_n}(t)$ converge pointwise to the characteristic function of P. Since we assume that $\lim_{n\to\infty} \phi_{a_n}(t) = \phi(t)$, we can conclude that $\phi(t)$ is the characteristic function of P.

We now demonstrate that the entire sequence $(P_n)_{n\geq 1}$ of probability measures weakly converges to P. Let $h(x)$ be a bounded and continuous function. We must show that

$$\lim_{n \to \infty} \int_{\mathbb{R}} h(x) \, dP_n(x) = \int_{\mathbb{R}} f(x) \, dP(x).$$

To prove this, we use the following fact from real analysis:

Given a numerical sequence $(\alpha_n)_{n \geq 1}$, if every subsequence of $(\alpha_n)_{n \geq 1}$ contains a further subsequence that is convergent to β, then the original sequence $(\alpha_n)_{n \geq 1}$ is convergent to β.

Let $i_1, i_2, i_3, \ldots$ be a subsequence of $\mathbb{N}$. We consider the subsequence

$$\left(\int h(x) \, dP_{i_j}(x) \right)_{j \geq 1}. \tag{14.8}$$

Since any subsequence of a tight sequence is tight, the sequence of probability measures $(P_{i_j})_{j \geq 1}$ is tight. By Prokhorov theorem, there exists a sub-subsequence i_{j_k}, where $j_1 < j_2 < j_3 < \cdots$, such that $F_{i_{j_k}}(x)$ converges pointwise to the distribution function $G(x)$ of a probability measure Q at every continuity point of $G(x)$. By Theorem 14.3, the characteristic functions associated to this sub-subsequence converge to the characteristic function of Q, which is also the characteristic function of P due to the assumption that $\lim_{k \to \infty} \phi_{i_{j_k}}(t) = \phi(t)$.

By the inversion formula in Theorem 9.5, Q and P have the same distribution. Hence,

$$\left(\int h(x) \, dP_{i_{j_k}}(x) \right)_{k \geq 1}$$

is a sub-subsequence of (14.8) converging to

$$\lim_{k \to \infty} \int h(x) \, dP_{i_{j_k}}(x) = \int h(x) \, dQ(x) = \int h(x) \, dP(x).$$

Therefore, $\int h \, dP_n$ converges to $\int h \, dP$ as $n \to \infty$, and this completes the proof of Lévy's continuity theorem.

We illustrate Lévy's continuity theorem with two classic examples. The first example involves discrete random variables converging to another discrete random variable, while the second example involves discrete random variables converging to a continuous random variable.

Example 14.3.1 (Binomial Converging to Poisson)
Let X_n be a random variable with binomial distribution $Bin(n, p_n)$ with $p_n = \lambda/n$, for a fixed positive constant λ. The characteristic function of X_n is

$$\phi_{X_n}(t) = (1 - p_n + p_n e^{it})^n.$$

If we take $n \to \infty$,

$$\lim_{n \to \infty} \phi_{X_n}(t) = \lim_{n \to \infty} \left(1 - \frac{\lambda}{n}(1 - e^{it}) \right)^n = e^{\lambda(e^{it} - 1)},$$

which is the characteristic function of a Poisson random variable. By Theorem 14.1, X_n converges to a Poisson random variable with mean λ in distribution.

Example 14.3.2 (Geometric Converging to Exponential)
Let λ be a positive constant and X_n be geometrically distributed with success probability $p_n = \lambda/n$, for $n = 1, 2, 3, \ldots$ The characteristic function of X_n is

$$\phi_{X_n}(t) = \frac{p_n e^{it}}{(1 - (1 - p_n)e^{it})}.$$

Let $Y_n = X_n/n$. We can check that $\phi_{Y_n}(t) = \phi_{X_n}(t/n)$ converges pointwise as $n \to \infty$,

$$\lim_{n\to\infty} \phi_{Y_n}(t) = \lim_{n\to\infty} \frac{e^{it/n}\lambda/n}{1 - (1 - \lambda/n)e^{it/n}} = \frac{\lambda}{\lambda - it}.$$

By Theorem 14.1, we have weak convergence $Y_n \xrightarrow{D} \text{Exp}(\lambda)$.

14.4   Central Limit Theorems

We use Levy's continuity theorem to prove a basic version of central limit theorem, which assumes i.i.d. random variables and finite variance.

Theorem 14.7 (Lindeberg–Lévy Central Limit Theorem)
Let $(X_k)_{k=1}^{\infty}$ be a sequence of i.i.d. random variables with finite mean $E[X_1] = \mu$ and finite variance $\text{Var}(X_1) = \sigma^2$. Then

$$\frac{\sum_{k=1}^{n} X_k - n\mu}{\sigma\sqrt{n}} \quad \text{converges in distribution to } N(0, 1).$$

Sketch of Proof We may assume $\mu = 0$ by re-defining $X_1 := X_1 - \mu$. Let $S_n \triangleq X_1 + X_2 + \cdots + X_n$.

Step 1. [4, Theorem 3.3.20] Since X_1 has variance σ^2, the characteristic function of X_1 is

$$\phi_{X_1}(t) = 1 - \frac{\sigma^2 t^2}{2} + o(t^2),$$

where $o(t^2)$ refers to a function $h(t)$ such that $\lim_{t\to 0} \frac{h(t)}{t^2} = 0$.

Step 2. Because the random variables X_k's are assumed to be independent, the characteristic function of $\frac{S_n}{\sigma\sqrt{n}}$ is

$$\left(1 - \frac{t^2}{2n} + o\Big(\frac{1}{n}\Big)\right)^n.$$

It can be written as $\left(1 + \frac{c_n}{n}\right)^n$, where c_n be the constant defined by

$$\frac{c_n}{n} = -\frac{t^2}{2n} + o\Big(\frac{1}{n}\Big).$$

Step 3. [4, Theorem 3.4.2] If $c_1, c_2, c_3, \ldots$ are complex numbers with $\lim_{n \to \infty} c_n = c$, then

$$\lim_{n \to \infty} (1 + c_n/n)^n = e^c.$$

Step 4. Because $c_n \to -\frac{t^2}{2} + o(1)$, we get $\lim_{n \to \infty} c_n = -t^2/2$. The characteristic function of $\frac{S_n}{\sigma \sqrt{n}}$ approaches $e^{-t^2/2}$ as $n \to \infty$. Since $e^{-t^2/2}$ is a continuous function, by Lévy's continuity theorem, $\frac{S_n}{\sigma \sqrt{n}}$ converges to the distribution whose characteristic function is $e^{-t^2/2}$.

Step 5. Steps 1–4 hold for any i.i.d. sequence with finite mean and variance. In particular, it is true if the random variables are i.i.d. standard Gaussian random variables. We use the fact that the sum of Gaussian random variables are Gaussian. When Steps 1 and 4 are applied to i.i.d. standard Gaussian random variables, the random variables $\frac{S_n}{\sigma \sqrt{n}}$ have standard Gaussian distribution for all n and hence must converge in distribution to $N(0, 1)$. The characteristic function of $N(0, 1)$ must be equal to $e^{-t^2/2}$. By the uniqueness of characteristic function, $\frac{S_n}{\sigma \sqrt{n}}$ converges in distribution to $N(0, 1)$. ∎

Based on the central limit theorem, we have the following approximations of binomial and Poisson distributions using normal distribution.

Example 14.4.1 (Normal Approximation of Binomial Distribution)
Suppose X_n's are i.i.d. Bernoulli random variables with success probability p. The mean and variance of X_n are p and $p(1-p)$, respectively. The sum $S_n = X_1 + X_2 + \cdots + X_n$ has distribution Binom(n, p). By the central limit theorem,

$$\frac{S_n - np}{\sqrt{np(1-p)}} \xrightarrow{D} N(0, 1).$$

Example 14.4.2 (Normal Approximation of Poisson Distribution)
We use the fact that the sum of n i.i.d. Poisson random variables with mean 1 is a Poisson random variable with mean n. Let S_n denote a Poisson random variable with mean n. Since the variance of a Poisson random variable with mean 1 is 1, we can apply the central limit theorem to get $(S_n - n)/\sqrt{n} \xrightarrow{D} N(0, 1).$

While the Lindeberg–Lévy central limit theorem is a useful result, it is limited in that it requires the random variables to be i.i.d. However, it is possible to relax the i.i.d. assumption and still obtain similar results. In the followings, we will state a central limit theorem for triangular array.

For each positive integer n, let k_n be a positive integer and let X_{n1}, $X_{n2}, \ldots, X_{n,k_n}$ be independent random variables defined on a common probability space. We can represent these random variables as a collection of rows, where the n-th row contains k_n random variables.

$$X_{11}, X_{12}, \ldots, X_{1,k_1}$$

$$X_{21}, X_{22}, X_{23}, \ldots, X_{2,k_2}$$

$$X_{31}, X_{32}, X_{33}, X_{34}, \ldots, X_{3,k_3}$$

$$\vdots$$

Note that the probability spaces may be different for different rows.

Suppose that the expectations of all random variables are zero, for notational simplicity, and let the variance of $X_{n,i}$ be σ_{ni}^2, for $n \geq 1$ and $1 \leq i \leq n$. We let S_n be the sum of the k_n random variables in the n-th row, and $s_n^2 = \sum_{i=1}^{k_n} \sigma_{ni}^2$, for $n \geq 1$, be the variance of S_n. We assume that $s_n^2 \to \infty$ as $n \to \infty$.

Theorem 14.8 (Central Limit Theorem for Triangular Array)
With the notation above, we have

$$\frac{X_{n1} + X_{n2} + \cdots + X_{n,k_n}}{s_n} \xrightarrow{D} N(0, 1)$$

if one of the following conditions hold:

- *(Lindeberg condition) For all $\epsilon > 0$,*

$$\frac{1}{s_n^2} \sum_{i=1}^{k_n} E[X_{n,i}^2 \mathbf{1}_{\{|X_{n,i}| \geq \epsilon s_n\}}] \to 0 \ as \ n \to \infty.$$

- *(Lyapunov condition) There exists $\delta > 0$ such that*

$$\frac{1}{s_n^{2+\delta}} \sum_{i=1}^{n} E[|X_{n,i}|^{2+\delta}] \to 0 \ as \ n \to \infty.$$

In the Lindeberg condition, the expectation is the integral

$$\int_{\{|x|\ge \epsilon s_n\}} x^2 \, d\mu_{n,i}(x),$$

integrating the function x^2 in the range $|x| \ge \epsilon s_n$ with respect to the measure $\mu_{n,i}$ induced by $X_{n,i}$. Intuitively, this means that the tail of each X_k must decrease fast enough so that the sum of the tail parts of the variances does not contribute significantly to the overall variance of S_n. We remark that for independent random variables, the conditions in the Lindeberg–Lévy central limit theorem imply the Lindeberg condition.

On the other hand, the Lyapunov condition is a stronger condition than the Lindeberg condition. Specifically, the Lyapunov condition requires that the tail probability of each random variable X_k decays sufficient fast so that certain power law holds. This condition is useful in some situations where the Lindeberg condition is difficult to check.

The proof is beyond the scope of this book. We refer the readers to [2] and [4]. Instead of going through the proof, we give an example and demonstrate how to apply this theorem.

Example 14.4.3 (Asymptotic Normality in Coupon Collection Problem)

In the coupon collection problem, we consider the random experiment that repeatedly and independently draws one out of n different coupons, where n is a fixed parameter. Let $T_{n,m}$ denote the first time that exactly m distinct coupons have been collected. This random variable is the sum of m independent geometric-distributed random variables,

$$T_{n,m} = X_{n,0} + X_{n,1} + X_{n,2} + \cdots + X_{n,m-1}.$$

The success probability of random variable $X_{n,i}$ is $1 - \frac{i}{n}$, and $X_{n,0}$ is equal to 1 with probability 1. For $i \ge 1$, $X_{n,i}$ counts the number of additional coupons needed to be drawn in order to get a new one after i distinct coupons have been collected.

The pmf of the random variable $X_{n,i}$ is given by

$$P(X_{n,i} = j) = (1 - p_{n,i})^{j-1} p_{n,i}$$

for $j \ge 1$, where $p_{n,i} \triangleq 1 - i/n$ is the probability of success. By using the moment generating function,

$$E[e^{X_{n,i}t}] = \frac{p_{n,i} e^t}{1 - (1 - p_{n,i})e^t},$$

we can compute the mean and variance of $X_{n,i}$ as

$$E[X_{n,i}] = \frac{1}{p_{ni}}, \quad \mathrm{Var}(X_{n,i}^2) = \frac{1 - p_{n,i}}{p_{n,i}^2}.$$

Additionally, the centered fourth moment of $X_{n,i}$ is given by

$$E[(X_{n,i} - E[X_{n,i}])^4] = \frac{1}{p_{n,i}^4}(1 - p_{n,i})(p_{n,i}^2 - 9p_{n,i} + 9).$$

Suppose we increase n and m simultaneously such that the ratio m/n approaches $1/2$. For instance, we may take $m = \lfloor(n + 1)/2\rfloor$. We are interested in the asymptotic distribution of the random variable $T_{n,\lfloor(n+1)/2\rfloor}$.

We can write the random variables $X_{n,i}$, for $n \geq 2$ and $0 \leq i \leq \lfloor(n + 1)/2\rfloor$, as a triangular array. The random variables in each row are independent, but not identically distributed. Hence, we cannot apply Theorem 14.7. Instead, we will apply Theorem 14.8, with $\delta = 2$ in the Lyapunov condition. We obtain the following upper bound for the sum of fourth moments,

$$\frac{1}{s_n^4} \sum_{i=0}^{\lfloor(n+1)/2\rfloor} E[(X_{n,i} - E[X_{n,i}])^4)]) = \frac{1}{s_n^4} \sum_{i=0}^{\lfloor(n+1)/2\rfloor} \frac{1}{p_{n,i}^4}(1 - p_{n,i})(p_{n,i}^2 - 9p_{n,i} + 9)$$

$$\leq \frac{1}{s_n^4} \sum_{i=0}^{\lfloor(n+1)/2\rfloor} \frac{9}{p_{n,i}^4} = \frac{9}{s_n^4} \sum_{i=0}^{\lfloor(n+1)/2\rfloor} \frac{1}{(1 - i/n)^4}.$$

We approximate the sum by a Riemann integral,

$$\frac{1}{s_n^4} \sum_{i=0}^{\lfloor(n+1)/2\rfloor} E[(X_{n,i} - E[X_{n,i}])^4)]) \leq \frac{9n}{s_n^4} \int_0^{1/2} \frac{1}{(1 - x)^4} \, dx. \tag{14.9}$$

Because

$$s_n^2 = \sum_{i=0}^{\lfloor(n+1)/2\rfloor} \frac{i/n}{(1 - i/n)^2} \approx n \int_0^{1/2} \frac{1}{(1 - x)^2} \, dx = n(1 - \log(2)),$$

we see that right-hand side of (14.9) approaches 0 as n approaches infinity. This verifies the Lyapunov condition. By using approximations by Riemann integral, we can show that the mean of $E[T_{n,\lfloor(n+1)/2\rfloor}] \to n \log 2$. We apply Theorem 14.8 to conclude that, as n approaches infinity,

$$\frac{T_{n,\lfloor(n+1)/2\rfloor} - n \log 2}{\sqrt{n(1 - \log 2)}} \xrightarrow{D} N(0, 1).$$

We implement the coupon collection problem using the following Python program and generate the histogram of the number of coupons we draw until we obtain $n/2$ distinct coupons out of n coupons.

```
from random import randint
from numpy import log, sqrt
import matplotlib.pyplot as plt

def coupon_collection(n,m):
    L = []                        # intitialize to empty list
    while True:
        L.append(randint(1,n))    # generate a new coupon
        if len(set(L)) >= m:
            break                 # break if we have m distinct coupons
    return(len(L))

n = 100                 # there are n types of coupons
```

```
m = 50                  # stop when m distinct coupons are collected
K = 50000               # repeat the experiment K times
T = [coupon_collection(n,m) for j in range(K)]  # generate data

mu = n*log(2)
sigma2 = n*(1-log(2))
x_min , x_max = (50,100)  # range of plots
bins = [i for i in range(x_min,x_max)]  # histogram bins
plt.hist(T,bins=bins)     # plot histogram
X = [x for x in range(x_min,x_max)]
Y = [K/sqrt(2*pi*sigma2)*exp(-(x-mu)**2/(2*sigma2)) for x in X]
plt.plot(X,Y, 'k-')
plt.xlabel('Number of coupons drawn until we get 50 out of 100 coupons')
```

We plot the histograms for $n = 100$ and $n = 1000$. The figures also show the probability density function of the limiting Gaussian distribution as a reference. The simulation results confirm that the probability distribution of the number of draws until obtaining $n/2$ distinct coupons is well approximated by a Gaussian distribution.

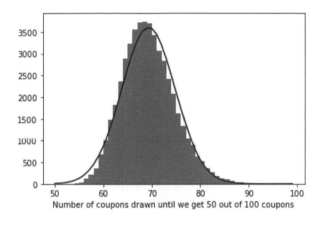

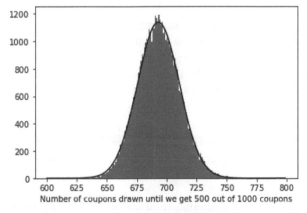

Problems

14.1. Continue with the probabilistic model of Polya's urn in Problem 13.11. For $i = 1, 2, 3, \ldots$, let X_i and Y_i, respectively, be numbers of white and black balls drawn in the first i steps. They satisfy $X_i \geq 0$, $Y_i \geq 0$, and $X_i + Y_i = i$ for all i:

(a) Show that, for $k = 1, 2, 3, \ldots, i$, the probability of $X_i = k$ is

$$P(X_i = k) = \binom{i}{k} \frac{w(w+1) \cdots (w+k-1)b(b+1) \cdots (b+i-k-1)}{(b+w)(b+w+1) \cdots (b+w+i-1)}.$$

(b) Hence, show that X_i / i converges in distribution to Beta(w, b).
   (Hint: Use Stirling's approximation and $\Gamma(x+a)/\Gamma(x+b) \approx x^{a-b}$.)

14.2. For $n = 1, 2, 3, \ldots$, let X_n denote a random variable with distribution $\Gamma(n\alpha, \beta)$. Prove that

$$\frac{1}{\sqrt{n\alpha\beta}} (X_n - n\alpha\beta) \xrightarrow{D} N(0, 1).$$

14.3. Let X_n be the random variable distributed according to the Gaussian distribution $N(0, n)$, for $n = 1, 2, 3, \ldots$. Show that this sequence of random variables fails to satisfy all four conditions in Lévy's continuity theorem.

14.4. For $n = 1, 2, 3, \ldots$, let X_n be a Gaussian random variable with distribution $N(0, 1/n)$. Show that X_n converges in distribution, and find the limit distribution.

14.5. Take $\mathbb{Z}$ as the sample space with the discrete topology $\mathcal{F} = 2^{\mathbb{Z}}$. Prove that a sequence of probability measures converges weakly if and only if it converges in total variation distance.

14.6. Given an arbitrary sequence of random variables $(X_n)_{n=1}^{\infty}$. Show that we can find positive constants c_n, for $n \geq 1$, such that $c_n X_n$ converges weakly to 0.

14.7. Let $(X_n)_{n=1}^{\infty}$ and $(Y_n)_{n=1}^{\infty}$ be two sequences of random variables such that X_n and Y_n are independent random variables defined on the same probability space. Show that if $X_n \xrightarrow{D} X$ and $Y_n \xrightarrow{D} Y$, then $X_n + Y_n$ converges to $X + Y$ in distribution (cf. Problem 10.10). (Hint: Apply Lévy's continuity theorem.)

14.8. Let $(X_n)_{n=1}^{\infty}$ be a sequence of random variables satisfying the following properties:

(i) For all n, the moment generating function $m_{X_n}(t)$ of X_n is defined for $t \in [-\delta, \delta]$.

(ii) For all $t \in [-\delta, \delta]$, the moment generating functions $m_{X_n}(t)$ converge to the moment generating function $m_X(t)$ of random variable X:
   (a) Prove that the sequence $(X_n)_{n \geq 1}$ is tight.
   (b) Prove that X_n converges to X in distribution.

14.9. (Continuous mapping theorem for convergence in distribution) Let X_n be a sequence of random variables converging to X in distribution. Let $f : \mathbb{R} \to \mathbb{R}$ be a function that is continuous at the set $S_f \subseteq \mathbb{R}$. Show that if $P(X \in S_f) = 1$, then $f(X_n) \xrightarrow{L} f(X)$. (Hint: Apply Lévy's continuity theorem.)

14.10. Consider a sequence of random variables $(X_n)_{n=1}^\infty$ that is uniformly bounded, i.e., there exists a constant c such that $|X_n| \leq c$ a.s. for all n. Let X be a random variable that is also bounded with probability 1. Prove that $X_n \xrightarrow{D} X$ if and only if $\lim_{n \to \infty} E[X_n^k] = E[X]$ for all $k \geq 1$. (Hint: Use the result in Problem 14.8.)

14.11. Generalize Example 14.4.3 to the case when m/n approaches a fixed constant r, for $0 < r < 1$.

14.12. (Uniform integrability) A sequence of random variables $(X_n)_{n=1}^\infty$ is said to be *uniformly integrable* if for every $\epsilon > 0$, there is a sufficiently large M such that $E[X_n \mathbf{1}_{|X_n| \geq M}] \leq \epsilon$ for all n. Prove the followings:

(a) A uniformly integrable sequence of random variables is tight.
(b) If a sequence of random variables is uniformly integrable and is converging in probability, then it is converging in the mean.

References

1. R.B. Ash, *Probability and Measure Theory*, 2nd edn. (Academic, San Diego, 2000)
2. P. Billingsley, *Probability and Measure*, 3rd edn. (Wiley, New York, 1995)
3. T.M. Cover, J.A. Thomas, *Elements of Information Theory*, 2nd edn. (Wiley, New Jersey, 2006)
4. R. Durrett, *Probability: Theory and Examples*, 5th edn.(Cambridge University Press, Cambridge, 2019)
5. J.M. Mendel, *Lessons in Estimation Theory for Signal Processing, Communications, and Control* (Prentice Hall, New Jersey, 1995)
6. G. Peyré, M. Cuturi, *Computational Optimal Transport – with Application to Data Science* (Now Publisher, Hanover, 2019)
7. W. Rudin, *Principles of Mathematical Analysis*, 3rd edn. (McGraw-Hill, New York, 1976)
8. W. Rudin, *Real and Complex Analysis*, 3rd edn. (McGraw-Hill, Singapore, 1986)
9. C. Villani, *Optimal Transport: Old and New*, Grundlehren der mathematischen Wissenschaften, vol. 338 (Springer, Berlin, 2009)
10. D. Williams, *Probability with Martingales* (Cambridge University Press, Cambridge, 1991)

© The Author(s), under exclusive license to Springer Nature Switzerland AG 2023   255
K. Shum, *Measure-Theoretic Probability*, Compact Textbooks in Mathematics,
https://doi.org/10.1007/978-3-031-49830-5

Index

A
Absolute continuity, 218
Algebra, 20
Almost everywhere, 111

B
Ball-and-bin model, 104
Borel algebra, 27
Borel–Cantelli lemma, 81, 82
Borel sets, 27
Borel zero-one law, 83

C
Cantor's diagonal argument, 18
Central limit theorem, 246
    Lindeberg condition, 248
    Lyapunov condition, 248
Change-of-variable formula, 123
Characteristic function, 153
Complete measure, 45
Conditional expectation
    continuous random variables, 225
    discrete random variables, 224
Continuous mapping theorem, 176
Convergence
    almost sure, 164
    in distribution, 168
    in the mean, 167
    in probability, 164
    sure, 164
    in total variation, 169
    weak, 238
Countable set, 17
Coupling, 132
    deterministic, 132
    inequality, 145
    maximal, 145

Coupon collection problem, 249
Covariance, 127

D
Data compression, 187
Distance function, 142
Distribution, 122
    Cantor, 7, 43
    chi, 126
    chi-square, 126
    continuous, 42
    Dirichlet, 4
    discrete, 43
    empirical, 169
    exponential, 121
    function, 36
    Gamma, 4
    Rayleigh, 122
    singular, 6
    uniform, 42
Dominated convergence theorem, 114

E
Equivalent random variables, 111
Event, 20
Expectation, 102
    conditioned on an event, 212
    conditioned on a partition, 213
    conditioned on a random variable, 217
    conditiond on a sub-sigma-algebra, 214

F
Fatou lemma, 113
Field, 20
Filtration, 228
Fubini theorem, 138

© The Editor(s) (if applicable) and The Author(s), under exclusive license to
Springer Nature Switzerland AG 2023
K. Shum, *Measure-Theoretic Probability*, Compact Textbooks in Mathematics,
https://doi.org/10.1007/978-3-031-49830-5

G
Glivenko–Cantelli theorem, 191

H
Hilbert space, 198

I
i.i.d., 81
Independence
　events, 71
　mutual, 78
　pairwise, 78
　random variables, 72
　σ-fields, 72
Inequality
　Cauchy–Schwarz, 181
　Chebyshev, 182
　Jensen, 183
　Markov, 166
　Minkowski, 197
Inner product of random variables, 196
Inverse transform method, 87
Inversion formula, 156
i.o., 81

K
Kantorovich's problem, 140
Kolmogorov zero-one law, 85
Kurtosis, 149

L
Lebesgue integral
　complex function, 99
　nonnegative function, 93
　real function, 97
　simple function, 90
Lévy's continuity theorem, 237
Limsup and liminf of sets, 22
Linear regression, 202, 205
L^1 norm, 109
LOTUS: law of the unconcious statistician, 125
Lower semi-continuity, 25
L^2 space, 195

M
Martingale, 228
　Doob's, 230
　stopping theorem, 232

Mathematical expectation, 102
Measurable function, 54
Measurable set, 55
Measurable space, 21
Measure
　Borel, 35
　counting, 24, 35
　Dirac, 42
　extension, 34
　extension theorem, 35
　function, 23
　Lebesgue, 42
　Lebesgue–Stieltjes, 37
　space, 23
MMSE estimator
　linear, 204
　nonlinear, 206
Moment generating function, 150
Moments, 149
Monge's problem, 139
Monotone convergence theorem, 95
Monte Carlo integration, 185

N
Normal number, 192
Null set, 45

O
Orthogonal principle, 203
Orthogonal random variables, 196

P
Parallelogram law, 201
π-λ theorem, 46
Polish space, 145
Polya's urn model, 236, 252
Pre-measure, 33
Projection function, 202
Projection theorem, 200
Prokhorov theorem, 244
Push-forward measure, 122

R
Radon–Nikodym theorem, 218
Random variable, 54
Random walk, 229
Riemann–Stieltjes integral, 9
Riesz–Fischer theorem, 198

S
Scheffé lemma, 129
σ-algebra, 21
σ-field, 21
σ-field generated by random variable, 72
σ-finite, 35
σ-subadditive, 38
Simple function, 90
Skewness, 149
Skorokhod's representation theorem, 172
Slutsky theorem, 179
Standard Borel space, 145
Standard deviation, 149
Stieltjes measure function, 36
Stopped process, 231
Stopping time, 230
Strong law of large numbers, 190
Support of a measure, 139

T
Tight sequence of measures, 241
Tonelli theorem, 137
Total variation distance, 142
Triangular array, 248

U
Uncorelated random variables, 127
Upper semi-continuity, 25

V
Vitali set, 29

W
Wasserstein distance, 142
Weak law of large numbers, 184, 185